T0121400

A Teacher's Guide to Bond-Type Triangles

A Visual Aid for Descriptive Inorganic Chemistry

William B. Jensen
Department of Chemistry
University of Cincinnati

2022

Order this book online at www.trafford.com
or email orders@trafford.com

Most Trafford titles are also available at major online book retailers.

Print information available on the last page.

ISBN: 978-1-6987-1318-2 (sc)
ISBN: 978-1-6987-1319-9 (e)

Library of Congress Control Number: 2022919746

Trafford rev. 10/25/2022

www.trafford.com

North America & international
toll-free: 844-688-6899 (USA & Canada)
fax: 812 355 4082

Dedicated to the Memory of
Hans August Georg Grimm
(1887-1958)
who started it all

Preface

This small monograph originated in a correspondence with Dr. Charles McCaw of Wincester College, Winchester, UK. Convinced of the pedagogical value of the bond-type triangle, Dr. McCaw was hoping to provide local teachers with supplementary materials illustrating its potential classroom applications and had contacted me concerning my various publications on this subject (1-5). I was, of course, more than to happy to provide him with copies of my published papers and reviews, with a list of additional references citing the triangle, and even with copies of several unpublished lectures that I had given on various occasions (6-12). Stimulated by the task of collecting these materials for my correspondent and infected by his obvious enthusiasm for this subject, it occurred to me that I might rework these various sources, as well as the large amount of as yet unpublished research in my files, into a small monograph on the bond-type triangle directed at teachers of introductory and inorganic chemistry.

In keeping with this goal and the title of this monograph, its purpose is to provide teachers with a background tutorial on the literature dealing with bond-type diagrams, as well as with a critical analysis of both the inherent limitations of such diagrams and their proper interpretation. My model in this endeavor is a small monograph for teachers published many years ago by S. Lewin on the nature and limitations of the solubility-product concept (13). Every teacher is aware that in presenting a subject to introductory students it is usually necessary to oversimplify and thus distort the true situation. The ability to minimize the negative consequences of such oversimplification is arguably a direct function of the depth of a teacher's understanding of what he or she is oversimplifying. The only tragedy is a teacher who distorts and oversimplifies without knowing that they are doing so. Hence the justification for short monographs of this sort specifically designed to provide teachers with critical background concerning concepts and models of potential pedagogical interest (14).

Though not explicitly directed at students and containing many details that would be inappropriate for an introductory course, this monograph also provides sufficient examples and applications to enable creative teachers to develop their own student-oriented teaching materials for this subject. I am, of course, fully aware that many of the problems and limitations of the solubility-product concept pointed out by Lewin nearly a half century ago continue to appear without correction in the textbook literature despite the fact that many of them have since been repeatedly reemphasized in the chemical education literature (15), a circumstance which lends a certain aura of futility to projects of this sort – but that, as they say, is unfortunately the nature of the beast.

References and Notes

1. W. B. Jensen, "The Historical Development of the van Arkel Bond-Type Triangle," *Bull. Hist. Chem.*, **1992-1993**, *13-14*, 47-59.

2. W. B. Jensen, "Quality versus Quantity," *Educ. Chem.*, **1994**, *31*, 10.

3. W. B. Jensen, "Bond Type Versus Structure Type," *Educ. Chem.*, **1994**, *31*, 94.

4. W. B. Jensen, "A Quantitative van Arkel Diagram," *J. Chem. Educ.*, **1995**, *72*, 395-398.

5. W. B. Jensen, "Logic, History, and the Chemistry Textbook: II. Can We Unmuddle the Chemistry Textbook?," *J. Chem. Educ.*, **1998**, *75*, 817-828

6. W. B. Jensen, "Bond-Type and Structure Sorting Maps," Symposium on Chemical Bonding, Department of Chemistry, University of Wisconsin, Madison, WI, July 1980.

7. W. B. Jensen, "Electronegativity and Bond Type," Invited Seminar, Department of Chemistry, Rochester Institute of Technology, Rochester, NY, 11 April, 1983.

8. W. B. Jensen, "A Short History of the van Arkel Bond-Type Triangle," Inorganic Seminar, Department of Chemistry, University of Cincinnati, Cincinnati, OH, 09

November, 1993.

9. W. B. Jensen, "The Historical Development of the van Arkel Bond-Type Triangle," 13th Biennial Conference on Chemical Education, Bucknell University, Lewisburg, PA, 31 July - 04 August 1994.

10. W. B. Jensen, "From van Arkel Triangles to Yeh Prisms: Bond-Type Diagrams and Structure Sorting Maps," Inorganic Seminar, Department of Chemistry, University of Cincinnati, Cincinnati, OH, 14 November 1995.

11. W. B. Jensen, "A History of the Chemical Bond and the van Arkel Bond-Type Triangle," Invited Seminar, Department of Chemistry, University of Arizona, Tucson, AZ, 31 October, 1996.

12. W. B. Jensen, "Bond-Type Triangles: An Overview," Invited Paper, 82nd Canadian Society of Chemistry Conference, Toronto, Canada, 30 May - 02 June 1999.

13. S. Lewin, *The Solubility Product Principle: An Introduction to its Uses and Limitations*, Interscience: New York, NY, 1960.

14. This was also presumably the motivation behind the excellent series of short specialty monographs published between the late 1950s and the 1980s by the Royal Institute (now the Royal Society) of Chemistry for science teachers, several of which also presented novel and innovative approaches to their subject matter. For an example of the latter, see P. A. H. Wyatt, *A Thermodynamic Bypass: Go to Log K*, Royal Society of Chemistry: London, 1982, Monograph No. 35.

15. See, for example, R. C. Goodman, R. H. Petrucci, "Is the Solubility Product Constant?," *J. Chem. Educ.*, **1965**, *42*, 104-105; L. Meites, J. S. F. Pode, H. C. Thomas, "Are Solubilities and Solubility Products Related?," *J. Chem. Educ.*, **1966**, *43*, 667-672, etc.

Table of Contents

I

Bond Types

1.1 Perspective

Perhaps the single greatest accomplishment of 20th-century chemistry was its successful development of the electronic theory of the chemical bond (1). Though recognition of the fact of chemical bonding dates back at least to the late 17th century and the first introduction of atomic or "corpuscular" representations of chemical interactions, early chemists could do little more than assign the phenomenon a name and study its molar consequences in the laboratory. This name of choice varied over time and included such terms as "attraction of composition," favored by such 18th-century chemists as Bergman and Fourcroy (2); "chemism," favored by some late 19th-century textbook writers (3), "chemical force," and "chemical affinity," which were perhaps the most popular of the various candidates.

Only with the discovery of the electrical nature of the atom at the turn of the 20th century was it possible to unambiguously identify chemical bonding with subatomic electrical forces of attraction and repulsion and to reduce all chemical change to the fundamental act of electron rearrangement. By 1913 it was further realized that this rearrangement could be associated with three distinct limiting mechanisms, corresponding to what are presently known as the ionic, covalent, and metallic bonding extremes.

1.2 The Ionic Bond

The electron-transfer mechanism for the polar or ionic bonding extreme was the first to receive recognition. Indeed, its essential characteristics were imperfectly anticipated as early as 1881 by the German physicist, Hermann von Helmholtz (4, 5). Arguing that Faraday's laws of electrolysis (1833) implied that electricity itself was particulate in nature, Helmholtz opted for a two-fluid model in which atoms of matter could combine with mobile particles of both positive and negative electricity. Neutral atoms were assumed to contain equal numbers of negative and positive electrical particles, whereas positive and negative ions were assumed to contain an excess of the corresponding electrical particle. Helmholtz further identified the number of excess electrical particles with the valence of the resulting ion, thus, in effect, postulating that all chemical combination was the result of the electrostatic attraction of oppositely charged ions. By making additional assumptions about average interatomic distances within molecules, he also demonstrated that the resulting electrostatic interactions between the component ions were of sufficient magnitude to account for the energy changes observed in typical chemical reactions.

With the exception of the British physicist, Sir Oliver Lodge, few physicists, and even fewer chemists, paid attention to Helmholtz's suggestions until they were revived by J. J. Thomson in conjunction with his ill-fated plum-pudding model of the atom in the period between 1904 and 1907 and reinterpreted in terms of a one-fluid model of electricity in which ionic charge was due to an excess or deficiency of a single mobile negative electrical particle or electron embedded in nonmobile sphere of positive electrification (6). Whereas Helmholtz had grafted his electrical particles onto an underlying substratum of classical Newtonian matter, Thomson had reduced matter itself to electricity.

In sharp contrast to the earlier reception accorded Helmholtz, Thomson's version of the polar or "electron transfer" model of bonding excited wide-spread enthusiasm and predictions of an impeding chemical revolution (7). In the United States it led to the development of a polar theory of organic reactivity in the period 1908-1925 in the hands of such chemists as Harry Shipley Fry, Lauder Jones, William A. Noyes, Julius Stieglitz, K. George Falk and John M. Nelson. In Germany, Richard Abegg (1904) successfully connected it with the periodic table, and in 1916 it was translated into the idiom of the Bohr-Rutherford quantized nuclear atom (1911, 1913) by German physicist, Walther Kossel (8).

Though crude quantitative calculations using the ionic model also date from the late 19th century (Richarz 1894, Ebert 1895), it was only with the development of quantitative lattice calculations for ionic crystals by Born and Landé in the period 1918-1919, and their graphical representation using thermochemical cycles (Haber 1919), that they truly became predictive (9, 10). Less rigorous results were also obtained in the field of coordination chemistry in the 1920s and 1930s (Kossel 1920, Magnus 1922, Garrick 1930), culminating in the formulation of the ionic radius-ratio rules (Hüttig 1920, Magnus 1922, Goldschmidt 1926) for the prediction of structure, and the Kapustinski equation (1933) for the rapid approximate calculation of complex lattice energies (11-15). Though a few useful extensions were still to come, by 1929 most of the basic components of our present-day model of the ionic bond were in place and were summarized that same year by the Dutch chemists, Anton Eduard van Arkel and Jan Hendrick de Boer, in their monograph, *Chemische Binding als Electrostatisch Verschijnsel (Chemical Bonding as an Electrostatic Phenomenon)* – a title which perfectly captured the newly won recognition that the previously mysterious force of "chemism" had been successfully reduced to a topic in applied electrostatics (16).

1.3 The Covalent Bond

Recognition of the electron-sharing mechanism for the homopolar or covalent bonding extreme also dates back to the first decade of the 20th century and the proposals of the German physicist, Johannes Stark (1908), and the German chemist, Hugo Kauffmann (1908). Related models were also suggested by William Ramsay (1908), Niels Bohr (1913), J. J. Thomson (1914), William C. Arsem (1914), Alfred Parsons (1915) and others (8). However, the overwhelming success of the ionic model and its rapid quantification tended to eclipse these electron-sharing models to such an extent that in 1913 G. N. Lewis felt compelled to write a paper arguing that the physical properties of typical organic compounds were incompatible with the ionic model and, indeed, strongly suggested the necessity of a second "nonpolar" bonding mechanism (17).

A successful candidate for this nonpolar mechanism was finally provided by Lewis himself in his famous 1916 paper on the electron-pair bond (18). This received widespread attention as a result of its extension and popularization by Irving Langmuir in the period 1919-1921 (19). Indeed, it was Langmuir who first introduced the term "covalent bond" as a replacement for the more cumbersome term "electron-pair bond" favored by Lewis. Beginning in the mid-1920s, qualitative extensions and applications of the covalent bond were also made in the field of organic chemistry by the British chemists, Arthur Lapworth, Robert Robinson, Thomas Lowry, and Christopher Ingold, and in the field of coordination chemistry by American chemist, Maurice Huggins, and the British chemist, Nevil Sidgwick (20, 21). This qualitative version of the covalent bond was first summarized in book form by Lewis in his 1923 monograph, *Valence and the Structure of Atoms and Molecules* (22), and in greater deal by Sidgwick in his 1927 monograph, *The Electronic Theory of Valency* (23).

Reconciliation of the shared electron-pair model with the newer quantum mechanics of Schrödinger and Heisenberg began two years before the appearance of the van Arkel - de Boer book on the ionic model with the publication of the classic 1927 paper on the H_2 molecule by the German physicists Walther Heitler and Fritz London (24). This work was further extended by Linus Pauling, using the concepts of orbital hybridization and resonance, in a series of papers published in the 1930s (25). Now known as the valence bond or VB method, it was first summarized in book form by Pauling in his 1939 monograph, *The Nature of the Chemical Bond* (26).

A second, more revolutionary, approach to covalent bonding, which evolved out of the study of molecular spectroscopy, was also developed about the same time (Hund 1926, Mulliken 1929, Lennard-Jones 1929) (27). Known as the molecular-orbital or MO method, because of its attempt to extend the delocalized orbital model of atomic spectra to polyatomic molecules, it was first successfully applied in a simplified quantitative form to unsaturated and aromatic species by the German physicist, Eric Hückel, in 1930 and was first summarized in book form in his 1938 monograph, *Grundzüge der Theorie ungesättigter und aromatischer Verbindungen* (28).

Initially most chemists tended to favor the VB approach, largely because its qualitative consequences could be represented using traditional Lewis dot diagrams. The MO approach, however, was better suited to the rationalization of spectra and was computationally more tractable. Consequently, with the widespread adoption of commercial IR and UV spectroscopy, the rise of organic photochemistry, and the increasing availability of computing facilities in the 1950s, both experimental and theoretical chemists began to increasingly favor the use of MO theory, which has essentially dominated the advanced chemical literature since the 1960s, though there have been some recent attempts to revive VB theory as well (29).

1.4 The Metallic Bond

Recognition of the free-electron mechanism for the metallic bonding extreme dates back to the work of the German physicist, Paul Drude, and the Dutch theorist, Hendrik Lorentz, at the turn of the 20th century (30, 31). Both assumed that the weakly bound conduction electrons in metals could be modeled using the classical kinetic theory of gases. Though this "electron-gas" model gave an adequate qualitative rationale of metallic properties, it incorrectly predicted both the heat capacity of metals and the temperature dependence of their electrical conductivity. In addition, it could not explain their magnetic properties (32). In 1928 the German physicist, Arnold Sommerfeld, partially resolved these problems by imposing quantum restrictions on the possible electron distributions – a lead further developed the same year by the Swiss physicist, Felix Block, in a classic paper which laid the foundations of modern band theory (33, 34).

The possible relevance of Drude's original free-electron model of metals to the theory of the chemical bond was first pointed out by G. N. Lewis in the same 1913 paper in which he had so forcefully argued for the separate existence of the nonpolar or covalent bond. In the final section of this paper, entitled "A Third Type of Chemical Bond," Lewis noted that (17):

To the polar and nonpolar types of chemical compound we may add a third, the metallic. In the first type the electrons occupy fixed positions within the atom. In the second type the electrons move freely from atom to atom within the molecule. In the third or metallic type the electron is free to move even outside the molecule.

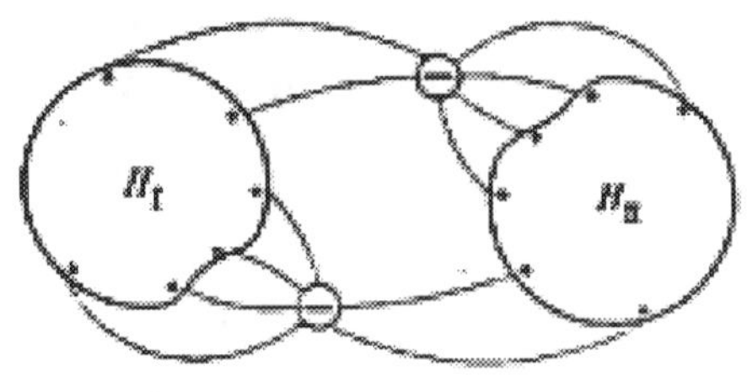

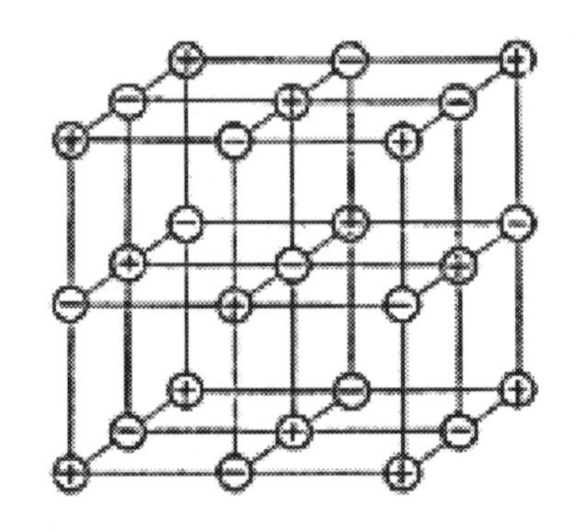

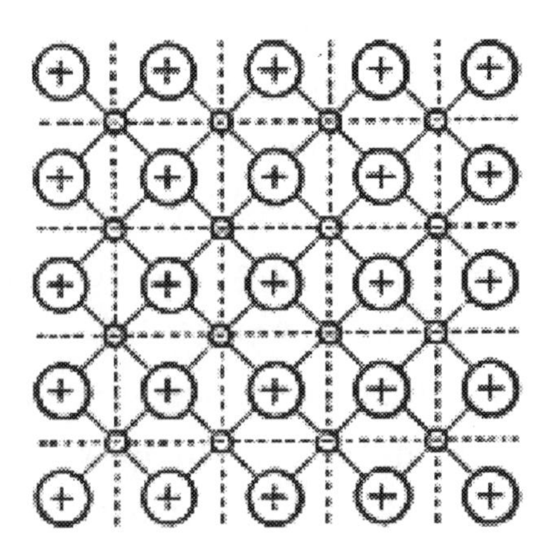

Two years later the German physicist, Johannes Stark, independently expressed the same idea and also made the first attempt to visualize all three bonding extremes (figure 1), though he pictured metals as a rigid lattice of positive ions and electrons rather than as a free-electron gas

Figure 1. Stark's 1915 depiction of the covalent bond in H_2 as two shared electrons (top), the ionic bond in NaCl as a lattice of positive and negative ions (center), and the metallic bond as a lattice of positive ions and free electrons (bottom).

(35) – a metallic model also advocated by the British physicist, Frederick A. Lindemann (36). Yet a second attempt to visualize Lewis's three bonding extremes, as well as weaker inter-molecular attractions, was made eight years later by the German chemist, Carl Angelo Knorr, using Bohr's dynamic atom model (figure 2) (37).

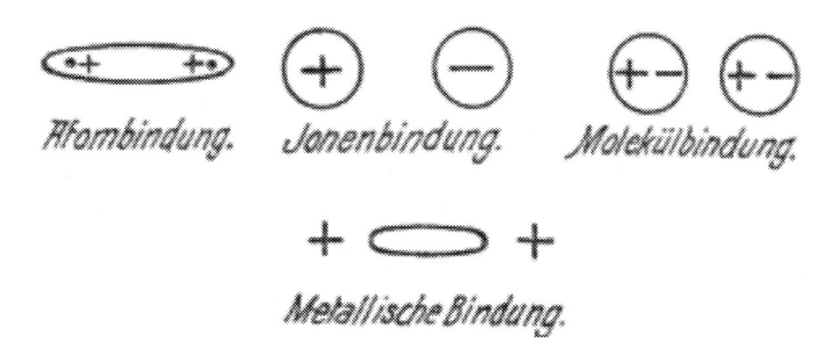

Figure 2. Knorr's 1923 depiction of the covalent (left), ionic (center) intermolecular (right) and metallic (bottom) bonds using Bohr's model of the atom.

The initial attraction of both the ionic and covalent models lay in their ability to qualitatively correlate the known compositions and structures of compounds with the number of valence electrons present in their component atoms. These "electron-count correlations" strongly appealed to chemists and are still the basis of much current chemical thought, as witnessed by the more recent development of valence-shell electron-pair repulsion theory (VSEPR) and the current rash of electron-counting rules for cluster species (38, 39). Only after these bonding models had proved capable of qualitatively correlating electron counts with composition and structure for significant classes of compounds did chemists exhibit a further interest in their quantification and in their ability to predict cohesive energies and other properties.

In sharp contrast, the history of the metallic bond has followed a very different scenario, since, to this day chemists have, with few exceptions, been unable to uncover a significant pattern governing the compositions and structures of intermetallic compounds and alloys (many of which are inherently nonstoichiometric), let alone establish simple electron-count correlations for them (40). As a consequence most mainstream chemists have displayed little interest in intermetallic compounds and the theoretical development of the metallic bonding model has been left largely to the solid-state physicist, who has, in turn, tended to stress the explication of thermal, electrical and optical properties, rather than cohesive energies and patterns of stoichiometry and structure. In addition, the resulting models have had a very different conceptual basis than those employed by the chemist and, even today, it is fair to say that such concepts as Brillouin zones and pseudopotentials are not part of the everyday working vocabulary of the average chemist.

Very few introductory chemistry texts make mention of the metallic bond and those that do still invoke the crude free-electron gas model of Drude to explain the high electrical conductivities of metals. If specific examples are cited, they invariably correspond to simple substances and all mention of the eccentricities of intermetallic compounds and alloys are carefully avoided. Indeed, it is fair to say that in English-speaking countries the systematic study of these compounds has never formed a major part of the mainstream chemical literature, having instead been largely consigned to the metallurgical literature. On the other hand, the same does not appear to be true of the German chemical literature, where a concerted effort to establish electron-counting correlations for intermetallic species has remained part of the province of the inorganic chemist, as exemplified by the significant contributions made by such chemists as Eduard Zintl and Ulrich Dehlinger throughout the 1930s (41). Perhaps one legacy of this difference in emphasis is the attack on the concept of the metallic bonding extreme made in the early 1990s by the American chemists, Leland Allen and the late Jeremy Burdett (42), and its subsequent successful defense by the German chemist, J. Christian Schön (43) – an attack which, in the present author's opinion, was based on a very basic misunderstanding of the nature and purpose of the bond-type triangles which are the main subject of this small monograph.

1.5 Intermediate Bond Types

The explicit recognition that chemical bonding could result from at least three distinct mechanisms for the requisite valence-electron rearrangement was coextensive with the further realization that these mechanisms corresponded to idealized limiting-case extremes and that many chemical bonds were actually intermediate in character. Thus Lewis followed his 1913 summary of the three types of chemical bond with the further statement that (17):

All known chemical compounds may be grouped in the three classes: nonpolar, polar and metallic; except in so far as the same compound may in part or at times fall under two of these groups.

Likewise Knorr, writing in 1923, also recognized the possibility of transitional bond types (including that between inter- and intramolecular bonding) and was able to further illustrate the various limiting-case extremes using the growing body of solid-state structural data that had been obtained from X-ray crystallography since the publication of Lewis's original paper (37):

These four extremely different bond types, between which there exist countless transitions and which can be schematically illustrated in the following manner [recall figure 2], also correspond to four different kinds of crystal lattice, namely the ionic lattice (cesium fluoride), the atom lattice (diamond), the molecular lattice (ice), and the metallic lattice (sodium).

Lewis later attempted to represent the gradual transition from idealized covalent bonding to idealized ionic bonding using his well-known electron dot notation (figure 3):

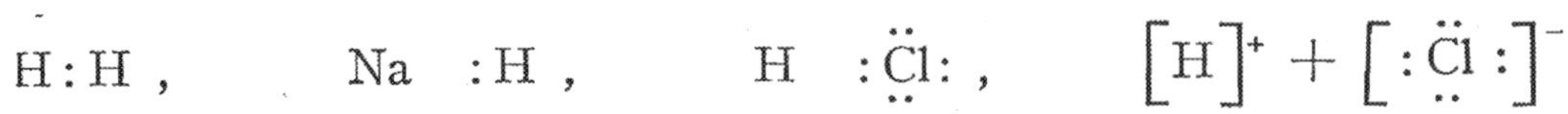

$$H:H\ ,\qquad Na\ \ :H\ ,\qquad H\ \ :\ddot{\underset{..}{Cl}}:\ ,\qquad [H]^{+}+[:\ddot{\underset{..}{Cl}}:]^{-}$$

Figure 3. Lewis' 1923 example of the transition between covalent and ionic bonding based on electronegativity differences and polar covalent bonds.

observing that (22):

The pair of electrons which constitutes the bond may lie between two atomic centers in such a position that there is no electric polarization, or it may be shifted toward one or another atom in order to give to that atom a negative, and consequently to the other atom a positive charge. But we can no longer speak of any atom as having an integral number of units of charge, except in the case where one atom takes exclusive possession of the bond pair, and forms an ion.

For example, we may suppose that the normal state of the hydrogen molecule is one in which the electron pair is symmetrically placed between the two atoms. In sodium hydride, on the other hand, we may regard the bonding pair as lying nearer to the hydrogen than to the sodium, making the hydrogen negative; while in hydrochloric acid the bond is shifted toward the chlorine, leaving the hydrogen with a positive charge. In the presence of a polar solvent, the chlorine assumes full possession of the bonding pair, and we have complete ionization.

Lewis further suggested that the resulting degree of unequal sharing or "polar covalent" bond formation, as it later came to called, was a function of the difference in the negative characters of the two interacting atoms, by which he meant the difference in their relative electronegativities (22):

Let us consider once for all that by a negative element or radical we mean one which tends to draw towards itself the electron pairs which constitute the outer shells of all neighboring atoms, and that an electropositive group is one that attracts to a less extent, or even repels, these electrons.

As is well known, nine years after Lewis wrote these lines, Pauling attempted to quantify these ideas by proposing both his famous thermochemical electronegativity scale and the existence of a direct correlation between the degree of bond polarity, on the one hand, and the difference in the relative electronegativities (ΔEN) of the two bonded atoms, on the other (26, 44). Thus we see that the Lewis-Pauling approach to bond polarity essentially took the idealized covalent bonding extreme as its reference state and discussed bond polarity as a deviation from this ideal which could be expressed in terms of the relative electronegativities of the two bonded atoms.

An alternative approach to polar covalency, due largely to the Polish chemist, Kasimir Fajans, was also proposed in the 1920s. This took instead the idealized ionic bonding extreme as its reference state and discussed bond polarity as a deviation from this ideal which could be expressed in terms of the polarizing ability of the cationic bond component, on the one hand, and the deformability or polarizability of the anionic bonding component, on the other – an idea summarized by Fajans in his Baker Lectures of 1930 (the same forum which would lead to the publication of Pauling's famous book nine years later) (45):

The deformation of the anions in the field of the cations ... lead, as can be readily seen [figure 4], to a partial return from the anion to the cation of the negative electricity transferred during the formation of the ions from neutral atoms. In this way the polar contrast between the ions is more or less strongly wiped out ... Since these deformation effects may be of very different degree from case to case, according to the size and charge of the cation and the deformability of the anion, it is plain that we have to expect a series of transitions between the extreme case of the linkage of rigid ions - the so-

called ideal ionic linkage – and the wholly nonpolar linkage.

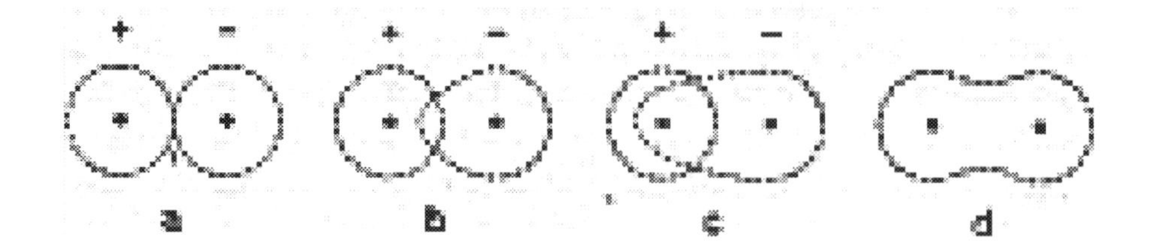

Figure 4. Fajan's 1930 depiction of the transition between ionic and covalent bonding via progressive ionic polarization.

During the first half of the 20th century, the Lewis-Pauling electronegativity approach tended to dominate the American and British chemical literature, whereas the Fajans ionic-polarization approach tended to dominate the continental European and Russian chemical literature – a regrettable bifurcation which has caused most chemists to overlook the fact that the various numerical scales of cationic polarizing ability, proposed over time within the context of the Fajans approach, are essentially identical to the various electronegativity scales which have been proposed over time within the context of the Lewis-Pauling approach. Thus the first general scale of cation polarizing ability was proposed by the American chemist, Groves Cartledge, in 1928, four years before Pauling proposed his thermochemical electronegativity scale (46-49). Known as the ionic potential (φ), it was defined as the charge (Z) to radius (r) ratio for the cation in question:

$$\varphi = (Z_{ion}/r_{ion}) \qquad [1]$$

In actual practice, however, Cartledge found that the square root of the ionic potential, rather than the ionic potential itself, gave better periodic trends and correlations with other atomic properties and, indeed, this parameter displays a linear correlation coefficient with modern *EN* scales of similar quality ($R = 0.96$) to those observed among the various alternative *EN* definitions themselves.

Similarly, in 1957 the Hungarian chemist, Bela Lakatos, proposed a definition of cation polarizing ability, which he called the "effective field strength," that was essentially identical to the well-known Allred-Rochow electrostatic force definition of electronegativity proposed one year later (50, 51):

$$EN = Z^*/r^2 = (Z - S)/r^2 \qquad [2]$$

where Z^* is the effective nuclear charge calculated using Slater's rules to estimate the value of the requisite screening constant (S).

Experimental studies of the gradual transition between ionic and metallic bonding were also undertaken in the 1930s, most notably by the German chemist, Eduard Zintl (41, 52), though there was no similar proposal of a simple parameter, similar to ΔEN, with which to characterize this transition. There are at least two reasons why the study of the progressive transitions between either the ionic or the covalent bonding extremes, on the one hand, and metallic bonding extreme, on the other, are more difficult to characterize than is the progressive transition between the covalent and ionic bonding extremes. The first of these has to do with the fact that few binary intermetallic compounds contain just one kind of bond. Thus the binary compounds studied by Zintl contained both polar A-B bonds and nonpolar B-B bonds and thus simultaneously represented both the transition from ionic to metallic bonding (the A-B bonds) and the transition from covalent to metallic bonding (the B-B bonds). Only the simple substances or elements seem to provide us with an unambiguous and direct transition between covalent and metallic bonding alone.

The second reason has to do with an inherent ambiguity in the theoretical treatment of covalent versus metallic bonding. In the above discussion it was implied that, just as the difference between the covalent and ionic bonding extremes lies in the distinction between the equal sharing of the valence electrons versus their complete transfer to the more electronegative atom, so the difference between the covalent and metallic bonding extremes lies in the distinction between localized versus delocalized sharing of the valence electrons. However, as most readers are aware, though the Lewis-VB approach to covalent bonding does indeed describe it in such a localized fashion, the alternative MO approach treats all chemical bonding as delocalized, even in the case of such molecules as methane (53).

The key to resolving this paradox lies in the realization that much of this so-called delocalization is an arbitrary artifact of the theoretical model and not necessarily an inherent physical feature of the molecules being described. In fact, it has been known since the 1950s that it is possible to transform the delocalized symmetry-adapted or spectroscopic orbitals of conventional MO theory into an alternative set of localized MOs which closely approximate the localized picture provided by the Lewis-VB model and that the degree of localization thus possible roughly varies from species to species in a manner consistent with our conventional chemical intuition concerning the varying degree of covalent versus metallic bonding character present (54). Numerous mathematical criteria for orbital localization have been explored since the 1960s, the most recent of which is the so-called "ELF" or "Electron Localization Function" of Savin and Nesper (55). When applied to a large range of species whose electron-density distributions have been calculated using extended Hückel MO theory, the resulting ELF diagrams largely mirror our traditional picture of how the valence-electron distribution changes on passing from typical ionic, to localized covalent, to delocalized metallic bonding situations.

1.6 References and Notes

1. This chapter is based on W. B. Jensen, "The Historical Development of the van Arkel Bond-Type Triangle," *Bull. Hist. Chem.*, **1992-1993**, *13-14*, 47-59, and W. B. Jensen, "The Origin of the Metallic Bond," *J. Chem. Educ.*, **2009**, *86*, 278-279.

2. For the term "attraction of composition," see T. Bergman, *A Dissertation on*

Elective Attractions, Murray: London, 1785, p. 5; and A. Fourcroy, *A General System of Chemical Knowledge*, Vol. 1, Cadell and Davies: London, 1804, pp. 94-121.

3. For the term "chemism," see G. F. Baker, *A Text-Book of Chemistry: Theoretic and Inorganic*, Morton: Louisville, KY, 1870, p. 2; and J. D. Steele, *A Popular Chemistry*, American Book Co: New York, NY, 1887, p. 2.

4. I begin with Helmholtz rather than with the earlier electrochemical theories of Davy (1807) and Berzelius (1812) because these theories have often been misrepresented by later historians. In actual fact, Berzelius did not postulate that chemical bonding was electrostatic. Rather he believed that free atoms contained net charges which were neutralized or destroyed on upon combination in order to produce caloric or heat (analogous to the spark produced in the discharge of a Leyden jar or capacitor). While Davy apparently believed that free atoms were neutral and only became electrified upon contact with one another, he did little more than state this belief and never developed it into a fully operative theory of chemical bonding.

5. H. von Helmholtz, "The Modern Development of Faraday's Conception of Electricity," *J. Chem. Soc.*, **1881**, *38*, 277-304. In actual fact, Helmholtz's suggestions merely replaced one puzzle with another. In the course of providing an electrostatic explanation of why atoms attracted one another to form molecules, he had created a substitute mystery – a new force responsible for the attraction between the underlying particles of matter and both the positive and the negative particles of electricity.

6. W. B. Jensen, "Abegg, Lewis, Langmuir and the Development of the Octet Rule," *J. Chem. Educ.*, **1984**, *61*, 191-200.

7. For a typical example of this somewhat premature enthusiasm, see R. K. Duncan, *The New Knowledge: A Simple Exposition of the New Physics and the New Chemistry in their Relation to the New Theory of Matter*, Barnes: New York, NY, 1908.

8. These proposals and extensions are discussed in great detail in A. N. Stranges, *Electrons and Valence: Development of the Theory, 1900-1925*, Texas A&M Press: College Station, TX, 1982. This is a well-documented history of the electronic theory of bonding up to 1925 with an excellent bibliography. The only point on which it is weak is in its coverage of the ionic model during the period 1916-1923, largely because of its over-emphasis on the emergence of Lewis's electron-pair bond.

9. These early calculations are summarized in S. Arrhenius, *Theories of Chemistry*, Longmans: London, 1907, pp. 61-64.

10. Summaries of the work of Born and Landé can be found reference 16, chapter 3, and in M. Born, *Atomtheorie des festen Zustandes*, Teubner: Leipzig, 1923. A more personal account is given in M. Born, *My Life: Recollections of a Nobel Laureate*, Scribner: New York, NY, 1975, pp. 181-183 and 188-190.

11. Summaries of the development of the electrostatic theory of coordination compounds can be found in reference 16, chapter 8, and in R. W. Parry, R. N. Keller, "Modern Developments: The Electrostatic Theory of Coordination Compounds," in J. C. Bailar, Ed., *The Chemistry of Coordination Compounds*, Reinhold: New York, NY, 1956, Chapter 3.

12. G. F. Hüttig, "Notiz zur Geometrie der Koordinationszahl," *Z. anorg. Chem.*, **1920**, *113*, 24-26.

13. A. Magnus, "Über chemische Komplexverbindungen," *Z. anorg. Chem.*, **1922**, *124*, 289-321.

14. V. M. Goldschmidt, *Geochemische Verteilungsgesetze der Elemente. VII. Die Gesetzte der Krystallochemie*, Dybwad: Oslo, 1926.

15. A. Kapustinsky, "Allgemeine Formel für die Gitterenergie von beliebiger Struktur," *Z. physik. Chem.*, **1933**, *22B*, 257-260.

16. A. E. van Arkel, J. H. de Boer, *Chemische Bindung als Electrostatisch Verschijnsel*, Centen: Amsterdam, 1929. This was translated into German as *Chemische Bindung als electrostatische Erscheinung*, Hirzel: Leipzig, 1931, and into French as *La valence et l'electrostatique*, Alcan: Paris, 1936. No English translation was ever made and all references are to the 1931 German edition. The first extensive English account of the literature dealing with the quantitative ionic model did not appear until 1932 and dealt only with the Born-Landé theory of ionic lattice energies – a small fraction of the material covered by the van Arkel - de Boer book. See J. Sherman, "Crystal Energies of Ionic Compounds and Thermochemical Applications," *Chem. Rev.*, **1931**, *11*, 93-170.

17. G. N. Lewis, "Valence and Tautomerism," *J. Am.. Chem. Soc.*, **1913**, *35*, 1148-1455.

18. G. N. Lewis, "The Atom and the Molecule," *J. Am.. Chem. Soc.*, **1916**, *38*, 762-785.

19. For the role of Langmuir, see references 6 and 8.

20. For a good account of the British School of organic reactivity, see W. H. Brock, *The Fontana History of Chemistry*, Fontana Press: London, 1992, Chapter 15.

21. For a summary of early applications to coordination chemistry, see R. W. Parry, R. N. Keller, "Modern Developments: The Electron-Pair Bond and Structure of Coordination Compounds," in J. C. Bailar, Ed., *The Chemistry of Coordination Compounds*, Reinhold: New York, NY, 1956, Chapter 4.

22. G. N. Lewis, *Valence and the Structure of Atoms and Molecules*, Chemical Catalog Co: New York, NY, 1923. This was translated into German as *Die Valenz und der Bau der Atome und Moleküle*, Vieweg: Braunschweig, 1927.

23. N. Sidgwick, *The Electronic Theory of Valency*, Oxford University Press: Oxford, 1927.

24. W. Heitler and F. London, "Wechselwirkung neutraler Atome und homöpolar Bindung nach der Quantenmechanik," *Zeit. Physik.*, **1927**, *44*, 455-472.

25. Many of Pauling's papers on VB theory are reprinted in B. Kamb et al., Eds., *Linus Pauling: Selected Scientific Papers*, 2 Vols., World Scientific Publishers: River Edge, NJ, 2001.

26. L. Pauling, *The Nature of the Chemical Bond*, Cornell University Press: Ithaca, NY, 1939.

27. Many of Mulliken's papers on MO theory are reprinted in D. A. Ramsay, J. Hinze, Eds., *Selected Papers of Robert S. Mulliken*, University of Chicago Press: Chicago, IL, 1975.

28. E. Hückel, *Grundzüge der Theorie ungesättiger und aromatischer Verbindungen*, Verlag Chemie: Berlin, 1938.

29. S. S. Shaik, P. C. Hiberty, *A Chemist's Guide to Valence-Bond Theory*, Wiley: Hoboken: NJ, 2007.

30. P. Drude, "Zur Electronentheorie der Metalle," *Ann. Phys.*, **1900**, *1*, 566-613; *Ibid.*, **1900**, *3*, 369-402; *Ibid*, **1902**, *7*, 687-692. Drude also references an even earlier free-electron model proposed by E. Reiche. An overview of this early work can also be found in K. Baedeker, *Die elektrischen Erscheinungen in metallischen Leitern,* Vieweg: Braunschweig, 1911.

31. H. A. Lorentz, *The Theory of Electrons and Its Application to Phenomena of*

Light and Radiant Heat, Teubner: Leipzig, 1906.

32. For an historical overview of the contribution of physicists to the theory of metals, see P. B. Allen, W. H. Butler, "Electrical Conduction in Metals," *Physics Today*, **1978**, *31 (12)*, 44-49.

33. A. Sommerfeld, "Zur Elektronentheorie der Metalle auf Grund der Fermischen Statik," *Z. Phys.*, **1928**, *47*, 1-32.

34. F. Block, "Über die Quantenmechanik der Elektronen in Kristallgittern," *Z. Phys.*, **1928**, *52*, 555-600.

35. J. Stark, *Prinzipien der Atomdynamik: Die Elekrizität im chemischen Atom*, Teil 3, Hirzel: Leipzig, 1915, pp. 120, 180, 193.

36. F. A. Lindemann, "Note on the Theory of the Metallic State," *Phil. Mag.*, **1915**, *29*, 127-140.

37. C. A. Knorr, "Eigenschaften chemischer Verbindungen und die Anordung der Elektronenbahen in ihren Molekülen," *Z. anorg. Chem.*, **1923**, *129*, 109-140. This was one of the first German papers to describe Lewis's electron-pair bond

38. For an overview of VSEPR, see R. J. Gillespie, I. Hargitai, *The VSEPR Model of Molecular Geometry*, Allyn and Bacon: Boston, 1991.

39. For an overview of electron-counting rules for clusters, see S. M. Owen, "Electron Counting in Clusters: A View of the Concepts," *Polyhedron*, **1988**, *7*, 253-283.

40. Electron-count correlations have been established for certain groups of intermetallic compounds and alloys, such as the well-known Hume-Rothery rules and the Engel-Brewer rules, but their range of application is quite limited.

41. For lead-in references see E. Zintl, "Intermetallische Verbindungen," *Angew. Chem.*, **1939**, *52*, 1-6; U. Dehlinger, "Die Chemie der metallischen Stoffe in Verhältnis zur klassischen Chemie," *Ibid.*, **1934**, *47*, 621-624; W. Klemm, "Intermetallische Verbindungen," *Ibid.*, **1935**, *48,* 713-723; and F. Weibke, "Intermetallische Verbindungen," *Z. Electrochem.*, **1938**, *44*, 209-221 and 263-282.

42. W. P. Anderson, J. K. Burdett, P. T. Czech, "What is the Metallic Bond?," *J. Am. Chem. Soc.*, **1994**, *116*, 8808-8809; L. C. Allen, J. F. Capitani, "What is the Metallic Bond?," *Ibid.*, **1994**, *116*, 8810; L. C. Allen, J. K. Burdett, "The Metallic Bond – Dead or Alive?," *Angew. Chem. Int. Ed. Engl.*, **1995**, *34*, 2003.

43. J. C. Schön, "Does the Death Knell Toll for the Metallic Bond?," *Angew. Chem. Int. Ed. Engl.*, **1995**, *34*, 1081-1083; J. C. Schön, "Reply," *Ibid.*, **1995**, *34*, 2004.

44. L. Pauling, "The Nature of the Chemical Bond: IV. The Energies of Single Bonds and the Relative Electronegativities of Atoms," *J. Am. Chem. Soc.*, **1932**, *54*, 3570-3582.

45. K. Fajans, *Radioelements and Isotopes; Chemical Forces and Optical Properties of Substances*, McGraw-Hill: New York, NY, 1931, pp. 1-21, 47-91. See also reference 16, chapters 4-6.

46. G. H. Cartledge, "Studies on the Periodic System. I. The Ionic Potential," *J. Am. Chem. Soc.*, **1928**, *50*, 2855-2863.

47. G. H. Cartledge, "Studies on the Periodic System. II. The Ionic Potential and Related Properties," *J. Am. Chem. Soc.*, **1928**, *50*, 2863-2872.

48. G. H. Cartledge, "Studies on the Periodic System. III. The Relation Between Ionizing Potentials and Ionic Potential," *J. Am. Chem. Soc.*, **1930**, *52*, 3076-3083.

49. G. H. Cartledge, "The Correlation of Thermochemical Data by the Ionic Potential," *J. Phys. Chem.*, **1951**, *55*, 248-256.

50. B. Lakatos, "Ein neuer Weg zum Berechnung des Polaritätsgrades der chemischen Binding," *Z. Elektrochem.*, **1957**, *61*, 942-949.

51. A. L. Allred, E. G. Rockow, "A Scale of Electronegativity Based on Electrostatic Force," *J. Inorg. Nucl. Chem.*, **1958**, *5*, 264-268.

52. Zintl's work is summarized and greatly extended in H. Schäfer, B. Eisenmann, W. Müller, "Zintl Phases; Transitions Between Metallic and Ionic Bonding," *Angew. Chem. Int. Ed. Engl.*, **1973**, *12*, 694-712.

53. W. A. Bernett, "Localized and Delocalized Molecular Orbital Descriptions of Methane," *J. Chem. Educ.*, **1969**, *46*, 746-749.

54. F. Weinhold, C. Landis, *Valency and Bonding: A Natural Bond Orbital Donor-Acceptor Approach*, Cambridge University Press: Cambridge, 2005, Chapter 5.

55. A. Savin, R. Nesper, S. Wengert, T. F. Fässler, "ELF: The Electron Localization Function," *Angew. Chem. Int. Ed. Engl.*, **1997**, *36*, 1809-1832.

II
Bond-Type Triangles

2.1 Right Triangles

The first attempt to construct a triangular bond-type diagram linking the three limiting-case bonding extremes appears to be that of the German chemist, Hans Georg Grimm (1). Beginning in 1928, he published a series of six papers dealing with the systematization and classification of binary compounds and simple substances (2-7). In order to trace out the pattern of ionic, covalent, and metallic bonding throughout the periodic table, Grimm constructed a series of both intra- and inter-row binary combination matrices for the elements, arranging them in order of increasing group number along the *x*-axis and decreasing group number along the *y*-axis. Since the noble-gas compounds were unknown at the time, this required that one consider only the seven chemically active groups of the short form of the periodic table and resulted in a 7x7 matrix which, after elimination of the redundant binary combinations, left a triangular array composed of 28 squares, each of which corresponded to either a real or potential binary compound or, in the case of the intra-row matrices, to a simple substance.

Using a characteristic cross-hatch scheme to indicate whether the resulting compound or simple substance was predominantly ionic, covalent or metallic in character, Grimm systematically constructed these triangular matrices for each possible combination of rows in the periodic table, finally concluding that they all corresponded to the same generalized pattern or "Dreieck-schema," which he summarized by means of the diagram shown at the top of the following page, in which ionic substances or "Salze" (S) were localized in the lower left-hand corner, covalent substances or "Atommoleküle" (A) were localized in the lower right-hand corner, and metallic substances or "Metalle" (M) were localized in the upper left-hand corner:

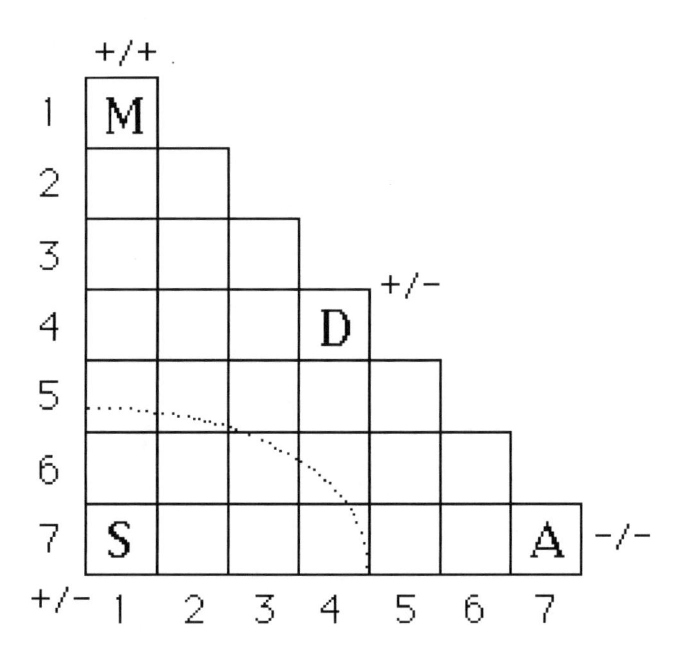

Also explicitly indicated on the diagram was the location of covalent solids or "Diamantartige Stoffe" (D) having infinitely polymerized 4/4 structures. Grimm did not intend this to represent yet a fourth bond-type. Rather he was highlighting his recent announcement of the rule that most covalent binary combinations (or Grimm-Sommerfeld compounds, as they are now commonly called in the solid-state literature) corresponding to this structure contained an average of four valence electrons per atom. In his triangular matrix, species having this particular average-electron count all fell along the diagonal connecting S and D, thus allowing him to illustrate how the transition from ionic to covalent solids for these species correlated with a change from the 8/8 and 6/6 crystal structures characteristic of the ionic examples to the 4/4 structures characteristic of the covalent examples.

The reason that Grimm had to construct so many separate diagrams was that he was lacking a quantitative electronegativity scale. Though the electronegativity concept had been introduced by Avogadro and Berzelius in the early 19th century, it was not fully quantified until the work of Pauling in 1932 (8). However, like all chemists since Berzelius, Grimm was fully aware of the qualitative fact that the electronegativity of the elements increased as the corresponding simple substances became less metallic and, like all chemists since Mendeleev and Lothar Meyer, he was also aware of the qualitative fact that the elements became increasingly electronegative as one moved from left to right across a given row of the periodic table (9). Thus, as long as he restricted each axis of his diagram to a single row, he was assured that the elements were placed in order of increasing electronegativity as a function of increasing group number, but in the absence of a single quantitative electronegativity scale he was unable to combine elements from different rows on the same axis, whence the necessity of his row by row survey. Once again, Grimm summarized these specific qualitative electronegativity trends on his generalized "Dreieckschema," using polar notation. Thus +/+ indicated that the metallic extreme involved the combination of two electropositive elements, -/- indicated that the covalent extreme involved the combination of two electronegative elements, and +/- indicated that the ionic extreme involved the combination of an electropositive with an electronegative element.

In 1936, the American chemist, Charles Stillwell, qualitatively arranged the main-block and transition elements in a single metallicity series roughly corresponding to the inverse of our current electronegativity scale and used this series to construct a single gigantic triangular matrix which simultaneously subsumed all of Grimm's smaller matrices (10, 11). Arranging the component elements in order of decreasing metallicity (i.e., increasing electronegativity) from bottom to top along the *y*-axis, and decreasing metallicity (i.e., increasing electronegativity) from left to right along the *x*-axis, Stillwell generated the following figure in which metallic combinations were localized in the lower left-hand corner, covalent combinations in the upper right-hand corner, and ionic combinations in the upper left-hand corner:

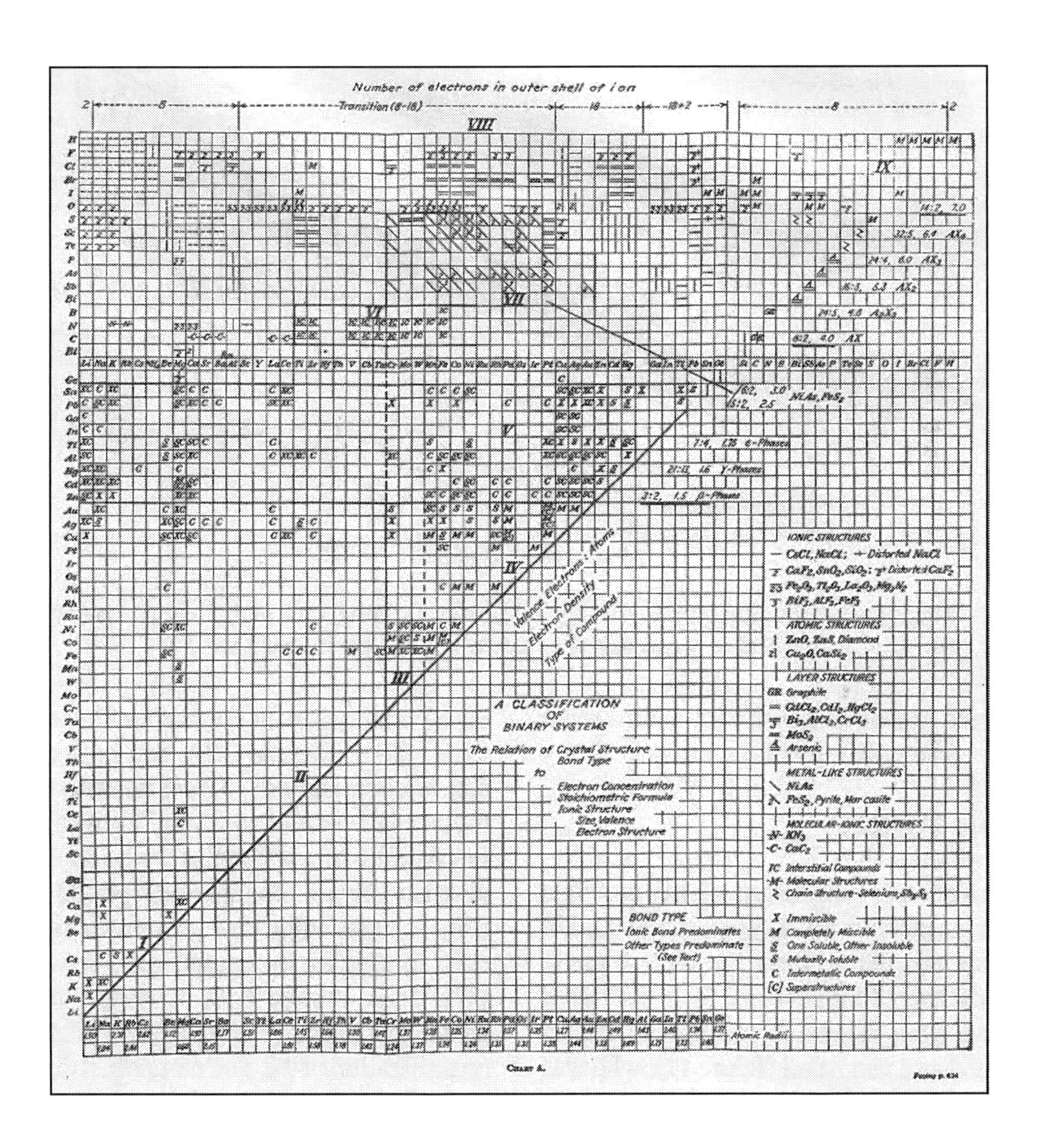

Stillwell was none too specific about the criteria used to construct his qualitative metallicity series, and close inspection shows many errors and inversions relative to our current electronegativity scales:

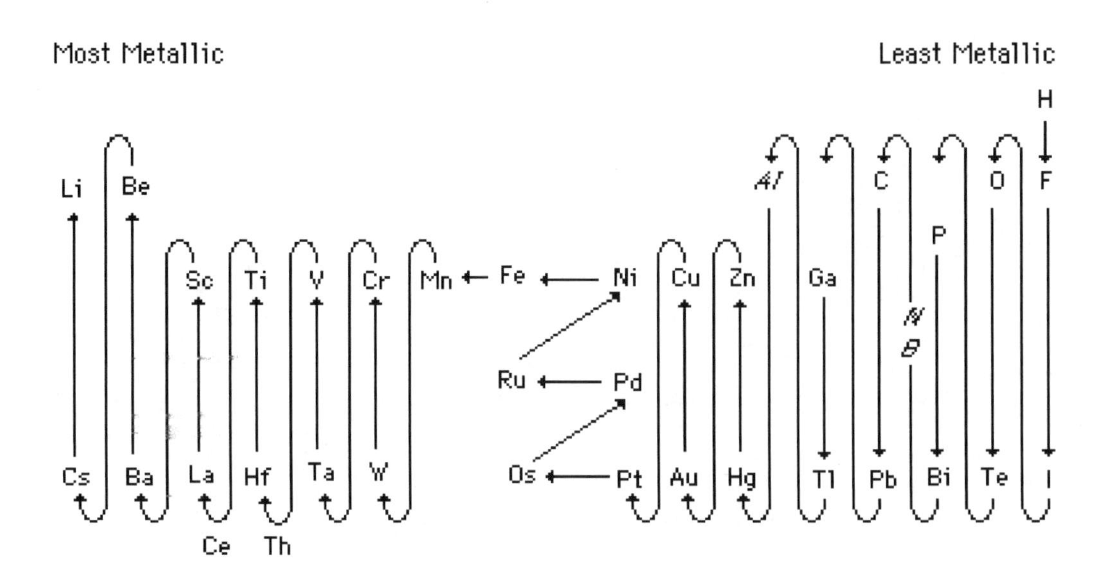

In addition, his primary purpose in constructing his diagram was to correlate the crystal structures of binary compounds with the electronic structures of their component atoms rather than with their bond types. This he did by filling the various squares with characteristic symbols and abbreviations denoting the structures of the compounds in question. Unfortunately, the resulting diagram was so cluttered and unwieldy that it failed to attract the attention of subsequent textbook authors. In keeping with his structural emphasis, it is also of interest to note that, unlike Grimm, Stillwell failed to explicitly label the corners of his triangle.

Twenty years after the publication of Stillwell's triangle, the Chinese chemist, Ping-Yuan Yeh, published a short note in the *Journal of Chemical Education* describing yet another version of the bond-type triangle (12). Arranging the component elements in order of increasing electronegativity from bottom to top along the *y*-axis, and from left to right along the *x*-axis, using the quantitative electronegativity values reported in Pauling's freshman textbook, *General Chemistry*, which Yeh was using in his introductory chemistry course, he obtained the diagram on the following page in which the metallic combinations were localized in the lower left-hand corner, the covalent combinations in the upper right-hand corner, and the

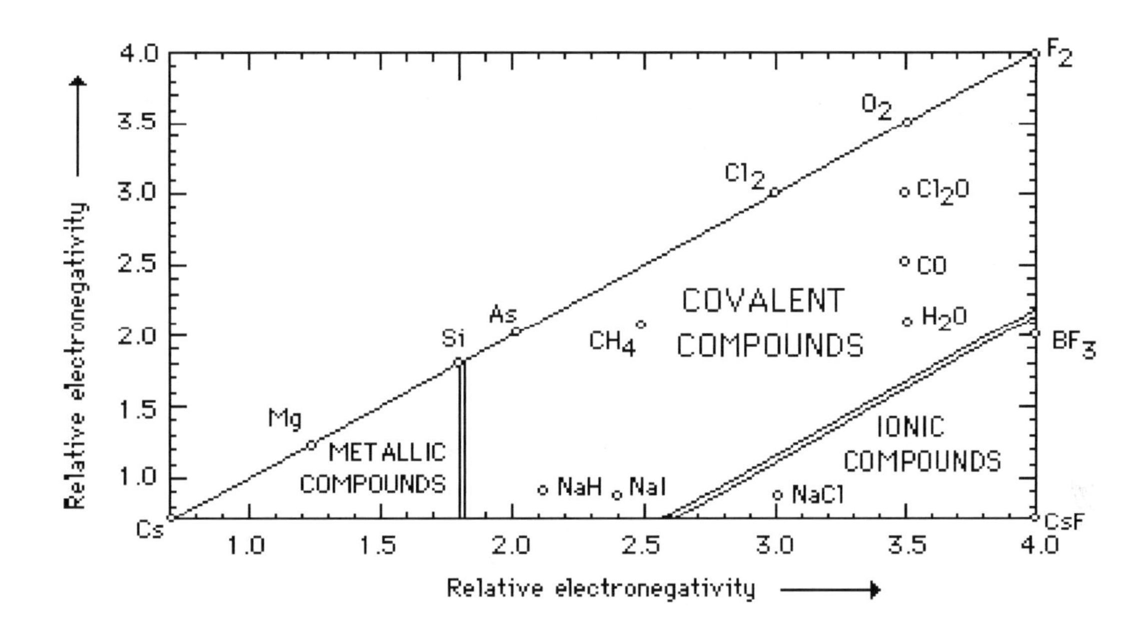

ionic combinations in the lower right-hand corner.

Like Stillwell, Yeh failed to explicitly label the corners of his triangle, and in keeping with the title of his note, "A Chart of Chemical Compounds Based on Electronegativities," he seems to have viewed his diagram more as a classification of compounds than as a classification of bond-types. The locations of only a few example binary compounds and simple substances were explicitly plotted on the diagram, all of them apparently taken from Pauling's text. Of particular note is the absence of any example intermetallic compounds. Indeed, a comprehensive application of this diagram would not have been possible in 1956 since Pauling's quantitative electronegativity scale was still incomplete. The electronegativity values for only 33 elements were reported in Pauling's freshman textbook – the same values that had appeared 17 years earlier in the first edition of his monograph, *The Nature of the Chemical Bond*, and it was not until the publication of the third edition of this book in 1960 that a complete scale was finally made available (13).

Yet a fourth variation of this bond-type triangle was published by the British chemist, N. W. Alcock, in his 1990 text, *Bonding and Structure* (14) in which the component elements were arranged in order of increasing electronegativity from top to bottom along the *y*-axis, and from left to right along the *x*-axis, and in which the metallic combinations were localized in the upper left-hand corner, the covalent combinations in the lower right-hand corner, and the ionic combinations in the upper right-hand corner:

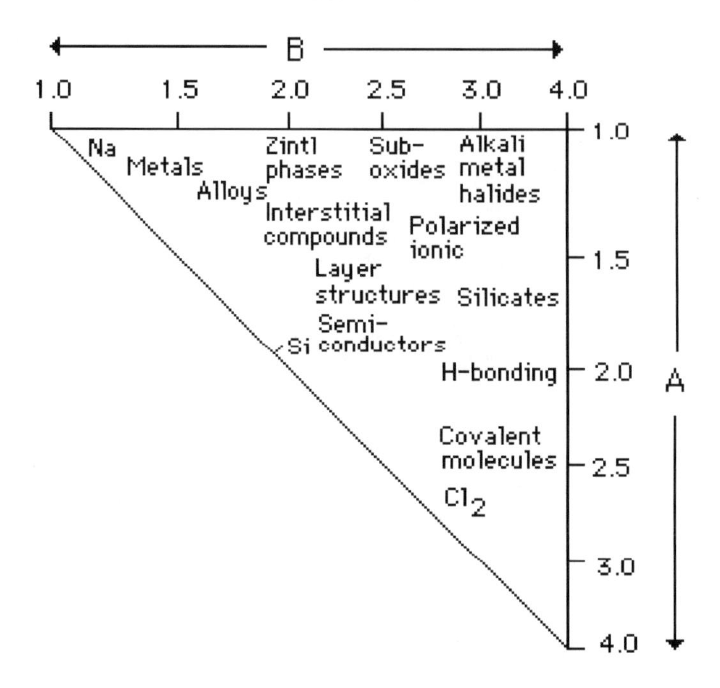

Despite the availability of complete quantitative electronegativity scales and the explicit placement of quantitative electronegativity values along both axes, Alcock did not plot the location of specific bonds or compounds within his triangle, but rather contented himself with indicating the approximate location of broad classes of compounds. In addition, like Stillwell and Yeh, he failed to explicitly label the corners of his triangle.

In summary, we see that between 1928 and 1990 at least four variations of an A-B bond-type triangle corresponding to various orientations of a right triangle were reported in the chemical literature, all of them based, either implicitly or explicitly, on a simple plot of the electronegativity of component A versus that of component B. As illustrated in the following summary diagrams, these various orientations were due to variations in which of the axes were used to plot the electronegativities (EN_A) of the cationic versus the anionic (EN_B) bond components and to whether the electronegativities in question were plotted in ascending or descending order (15):

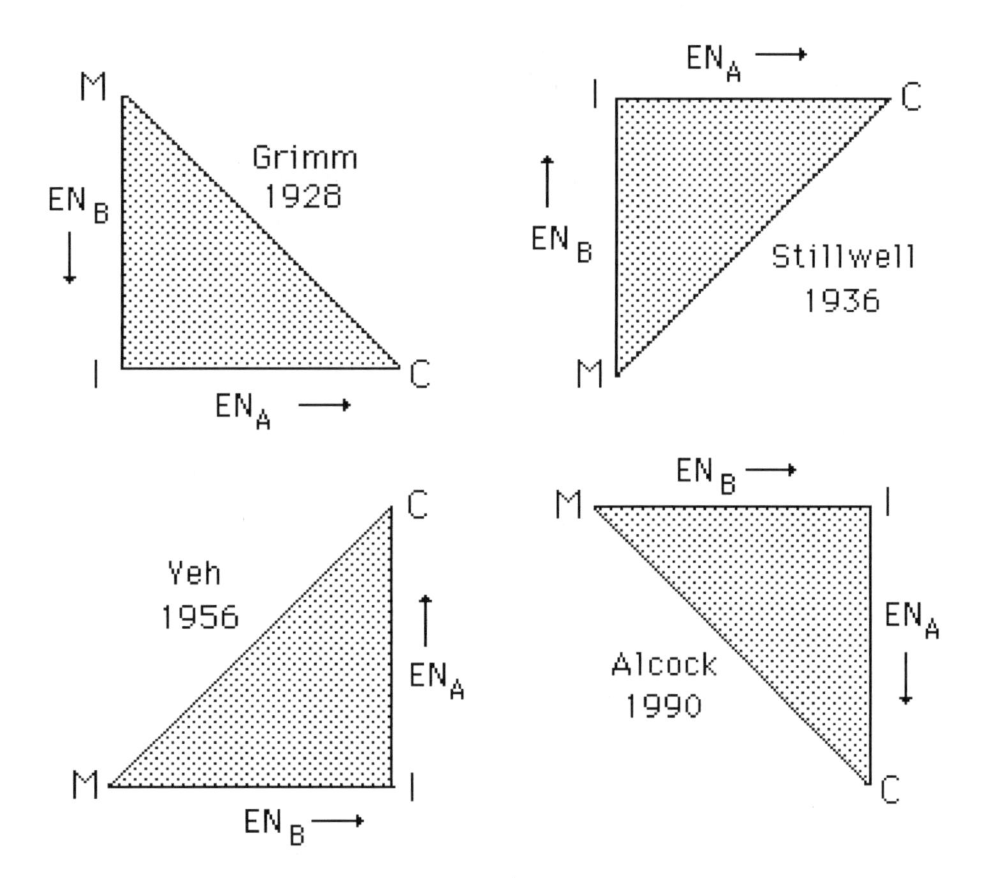

Aside from differences in orientation, the work of both Grimm and Stillwell preceded the quantified electronegativity scale available to Yeh and Alcock. Hence Grimm had to use group numbers and Stillwell an intuitive metallicity sequence as their coordinates – choices that are now known to roughly parallel changes in electronegativity.

2.2 Equilateral and Isosceles Triangles

Overlapping with the above developments, one also sees the gradual evolution of yet a second series of bond-type triangles based on the use of either equilateral or isosceles triangles rather than right triangles. Until quite recently these were purely qualitative in nature and, like the right-triangular diagrams discussed in the previous section, were often ambiguous when it came to the question of whether they were classifying compounds or their component bonds.

The first such qualitative diagram appears to have been introduced by the American chemists, Conard Fernelius and Richard Robey, in an article on "The Nature of the Metallic State" published in 1935 in the *Journal of Chemical Education* (16):

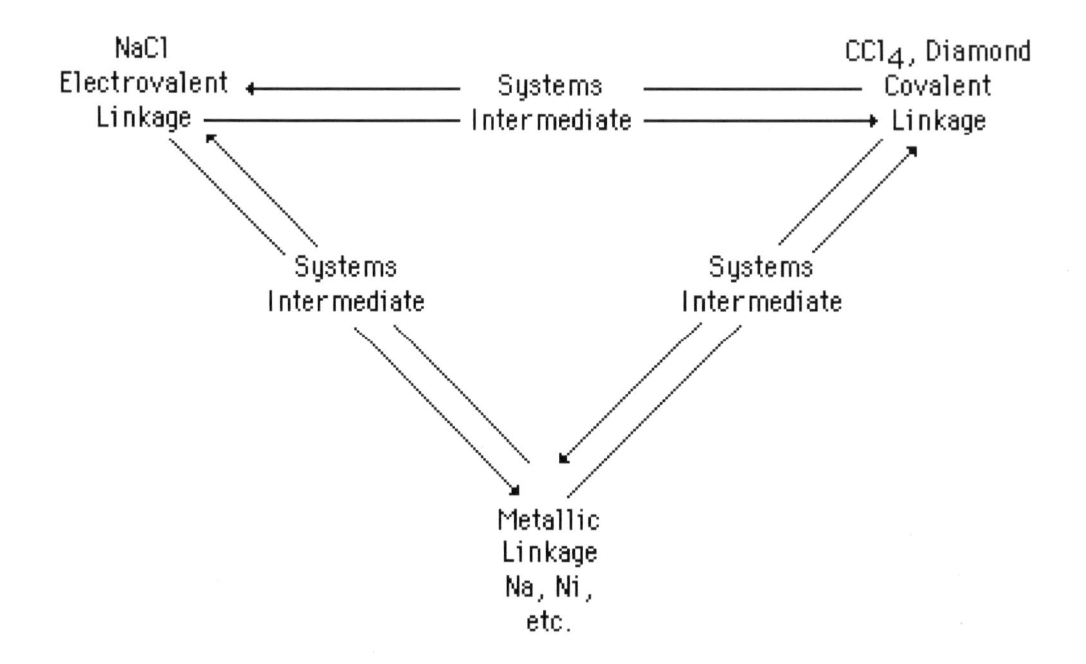

As may be seen, this diagram was unambiguously intended to deal with the classification of the bonds or "linkages" found within the few example substances also given on the diagram. In addition, the diagram explicitly indicated the possibility of mixed or intermediate bond types lying along each of the connecting edges, but made no provision for the possible existence of intermediate bonds lying within the body of the triangle and thus simultaneously partaking of all three limiting bond characteristics.

A very similar diagram was given six years later by the Dutch chemist, Anton van Arkel, in his monograph, *Molecules and Crystals* (17):

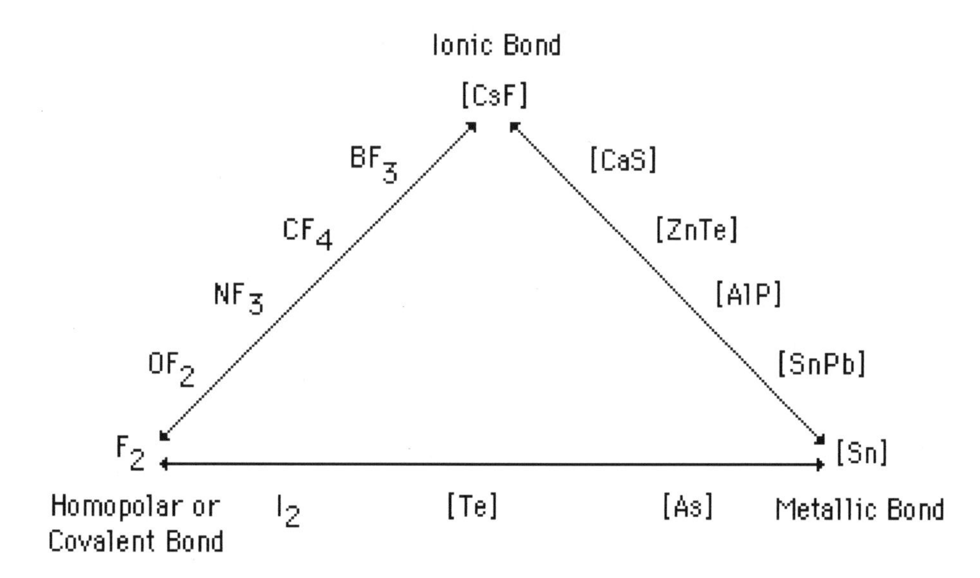

Once again the three vertices were explicitly labelled so as to indicate limiting bond type and example substances containing the bonds in question were arrayed at both the vertices and along the edges but not within the body of the triangle. As per the title of van Arkel's book, the example formulas lacking brackets denoted substance having discrete molecular structures, whereas those enclosed in square brackets denoted substances having infinitely polymerized crystal structures.

In 1947 van Arkel's colleague, the Dutch chemist, Jan Ketelaar, provided yet a third version of the bond-type diagram (first diagram on next page) and, for the first time, indicated the possibility of transitional bonds located with the body of the triangle as well as along the edges, though the significance of the broken horizontal lines was never explained (18).

The culminating development of the purely qualitative version of these diagrams was provided by William Jolly in the 1984 edition of his textbook, *Modern Inorganic Chemistry*, in which the entire body of the triangle was explicitly filled (albeit intuitively) with example compounds and simple

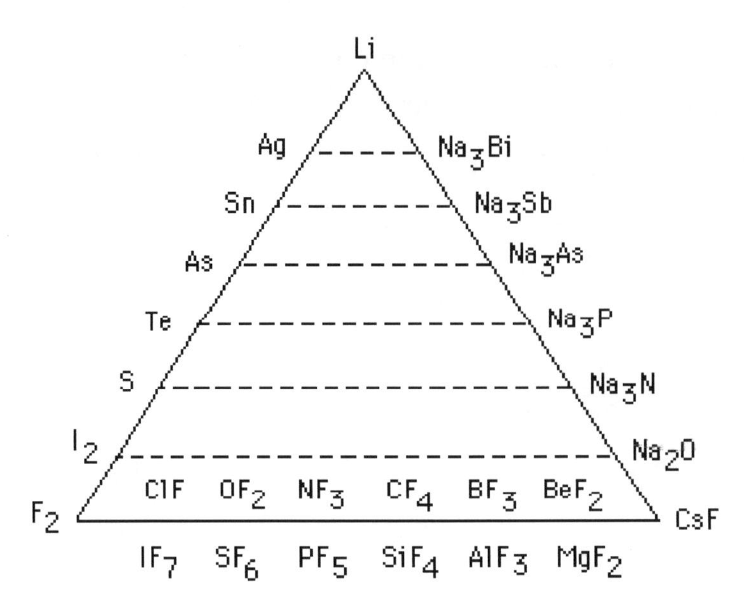

substances thought to contain the various intermediate and limiting cases bonds in question (19):

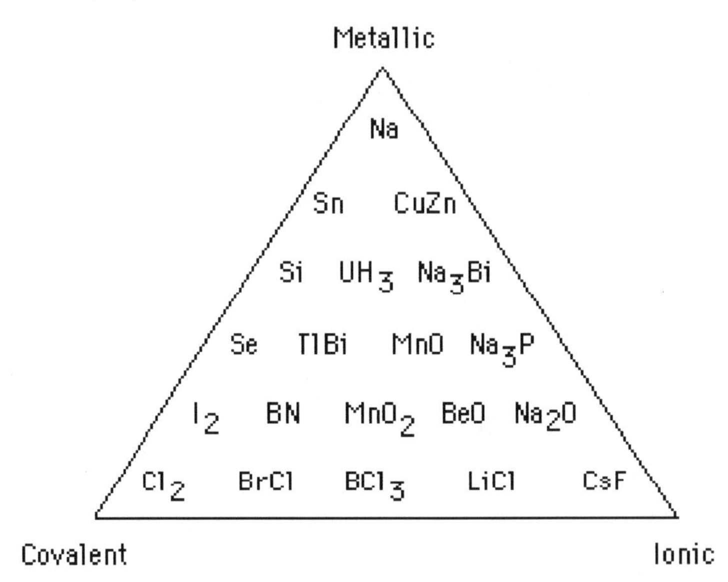

In all of the above diagrams the selection and placement of example compounds and simple substances within the triangle and on the edges was largely based on the chemical intuition of the author in question. The first attempt to make this placement process more explicitly systematic was made by Allen in 1992 (20). Illustrating his approach using the elements from the third row of the periodic table, he arranged them at equal intervals from left to right along the base of the triangle in order of their increasing group numbers. Dividing the body of the triangle into equally spaced horizontals, he then placed on the first horizontal the possible binary AB combinations formed by each adjacent pair of elements along the base, centering each half way between its components:

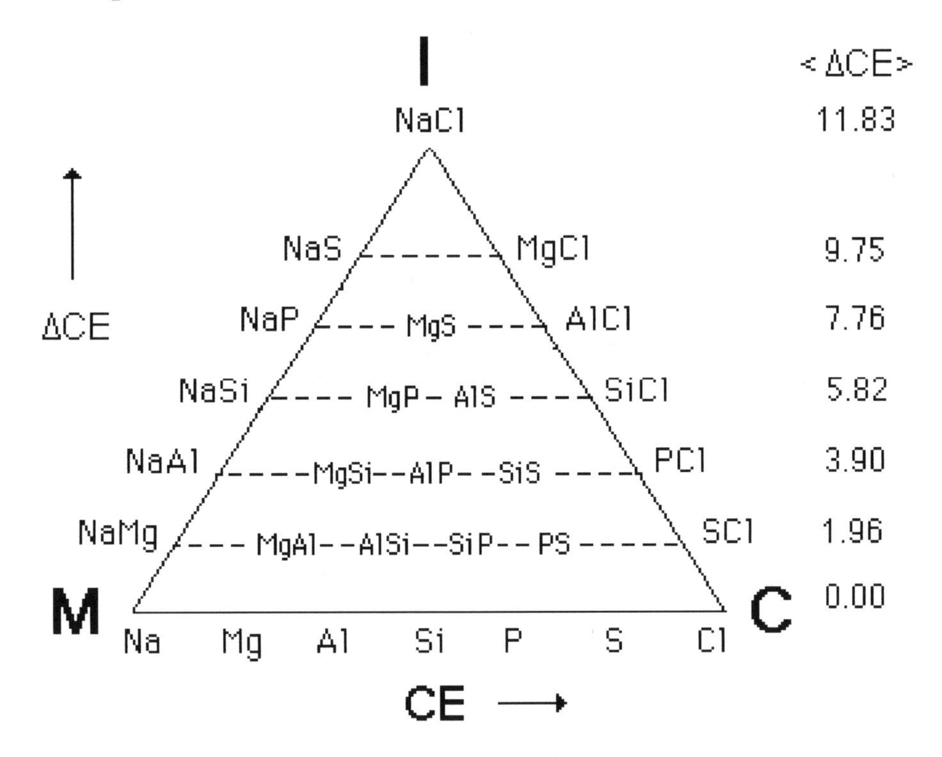

Thus MgAl appears on the first horizontal half way between Mg and Al on the base, AlSi appears half way between Al and Si, etc. On the second horizontal he placed the AB binary combinations created by taking the A component of the left binary on the first horizontal and the B component of the adjacent right binary, again centering them at equal intervals between their components on the previous horizontal. Thus MgP appears on the second horizontal half way between MgSi and AlP on the first horizontal, etc. This placement procedure was then repeated until one reached the top of the triangle.

Though Allen characterized his procedure as a "quantification of the van Arkel-Ketelaar triangle," it was really only semi-quantitative and dealt only with Ketelaar's version of the bond triangle. No coordinate system appeared on the diagram and all of the AB combinations in the triangle were instead qualitatively placed at equally spaced distances in the both the vertical and horizontal directions using the binary combination recipe just outlined. The label "$CE \rightarrow$" below the triangle is not a quantitative coordinate but was rather intended to emphasize that the elements were equally spaced along the base in qualitative order of their increasing electronegativities or configuration energies (*CE*), as Allen then preferred to call his personal version of the electronegativity scale (21). Likewise, the label "$\Delta CE \rightarrow$" to the left was intended to indicate that the average value of ΔCE for the binary combinations along each horizontal increased on moving from the bottom to the top of the triangle, as is apparent from the actual averages reported on the right. It was not intended to imply that the individual binaries had been plotted vertically in order of their individual ΔCE values, as is apparent from the actual ΔCE values for the various AB combinations found along the first horizontal:

NaMg (2.41), MgAl (1.89), AlSi (2.29), SiP (2.00) PS (2.10), SCl (1.66) [1]

These widely varying ΔCE values do, however, average to 1.96 as indicated on the right.

The first true quantification of the isosceles version of the bond-type triangle was independently proposed by both Jensen (1, 22-24) and Sproul (25-29) in the period 1992-1995. Both plotted the differences in the electronegativities of each binary AB combination versus the averages of their electronegativities. As shown below for Jensen's original triangle, the first of these parameters, termed the "bond ionicity" (*I*), was plotted in ascending order along the *y*-axis:

$$I = \Delta EN = (EN_B - EN_A) \qquad [2]$$

whereas the second parameter, termed the "bond covalency" (C), was plotted in ascending order along the x-axis:

$$C = EN_{av} = (EN_A + EN_B)/2 \qquad [3]$$

The resulting data points were found to naturally arrange themselves in a triangular pattern with the Cs-Cs bond at the metallic extreme in the lower left-hand corner, the F-F bond at the covalent extreme in the lower right-hand corner, and Cs-F bond at the ionic extreme at the upper apex:

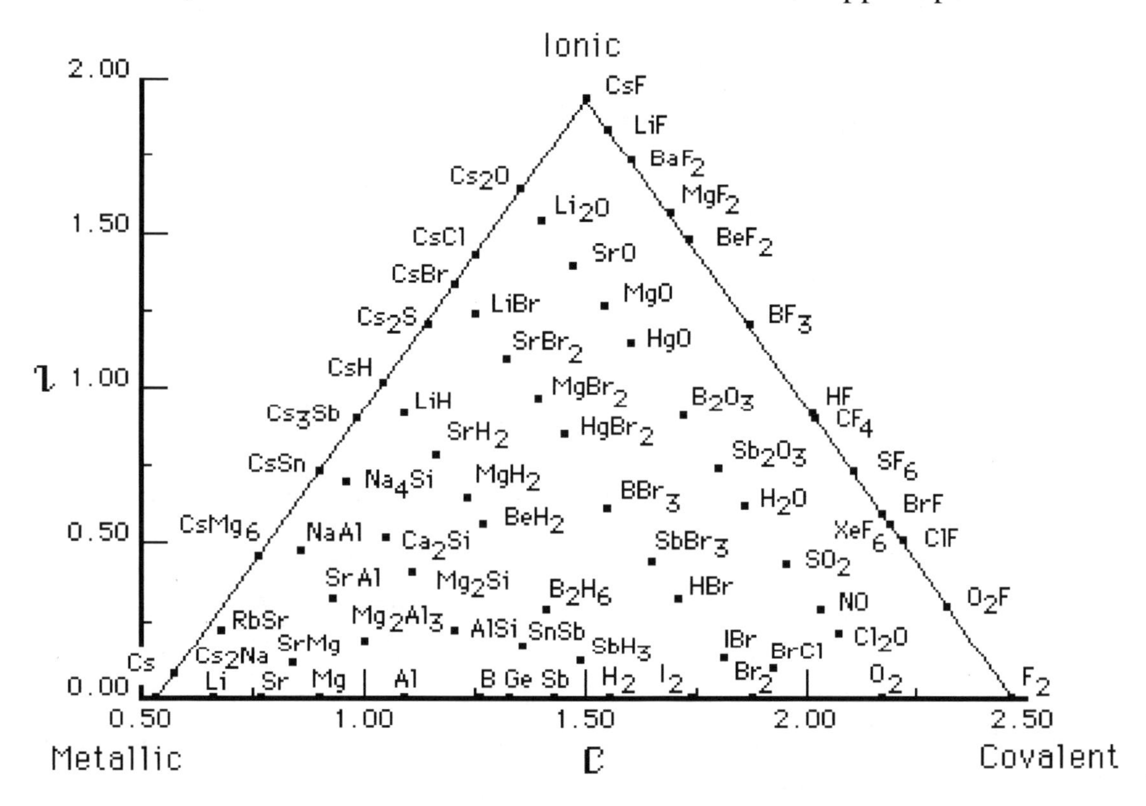

In summary, we see that between 1935 and 1995 at least six variations of an A-B bond-type triangle were reported in the chemical literature based on various orientations of either equilateral or isosceles triangles. In the absence of quantitative constraints, a wide variety of orientations were possible, as partially illustrated in the following diagrams. With quantification, however, the orientation has become standardized so as to conform to the version shown in lower right-hand corner:

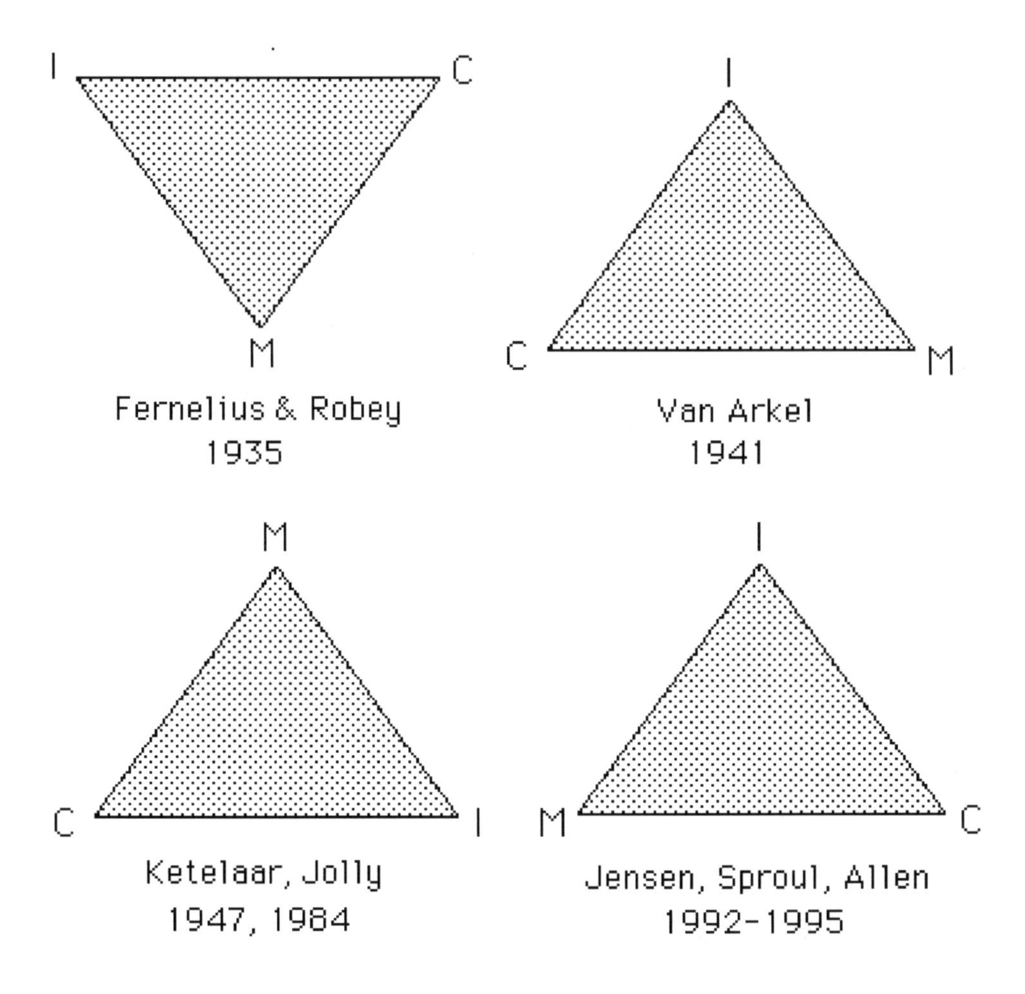

Once again, aside from differences in orientation, there were also some basic differences in the content of these various triangles. Thus Fernelius, and van Arkel not only relied on intuition to classify the bonds in their triangles, they only showed transitions along the triangle edges. Though Ketelarr and Jolly also showed transitions within the body of the triangle, their placements were once again based on intuition. It was Allen who finally provided a partial quantification of the Ketelaar triangle, but it was only with the independent work of Sproul and Jensen, that these types of triangles were finally fully quantified so as to account for bond transitions both within the triangle and along its edges.

2.3 Relative Advantages

The above survey has established the existence of basically two fundamental varieties of binary A-B bond-type triangles – one based on the use of a right triangle and a simple plot of EN_A versus EN_B, and the other based on the use of either an equilateral or isosceles triangle and a plot of ΔEN versus EN_{av}. For both historical reasons and ease of reference, we will, from this point on, refer to all right-angle diagrams, regardless of orientation, as Grimm triangles and all isosceles diagrams as Fernelius triangles, though the latter is often inaccurately referred to as a van Arkel-Ketelaar triangle by chemists (including the present author in his original papers) unfamiliar with the earlier work of Fernelius and Robey. In the next lecture we will explore how these two triangular diagrams are mathematically interrelated, but before doing so it is worth making a few preliminary observations about the relative merits of each.

The most obvious strength of the Grimm triangle is the simplicity of its coordinates – a feature which makes it an attractive candidate for use in an introductory chemistry course. I have been using it my own terminal course for health science majors for over a decade, where I gradually introduce it to students via a series of approximations. Starting with a crude division of simple substances into metals and nonmetals (illustrated with actual samples), I first construct the three-category classification for the resulting binary AB combinations:

	Metal	Nonmetal
Nonmetal		Covalent
Metal	Metallic	Ionic

I then refine the original classification of simple substances by pointing out the existence of the intermediate metalloids or semimetals, leading to the six-category elaboration of the above matrix:

	Metal	Semimetal	Nonmetal
Nonmetal			Covalent
Semimetal		Semi-Metallic	Polar Covalent
Metal	Metallic	Polar Metallic	Ionic

Lastly, I show how the simple substances corresponding to the main-block elements may be arranged in a single series, passing from metals through metalloids to nonmetals, as a function of the increasing electronegativity of their component atoms, and how this transforms the above matrix into a continuous plot of EN_A versus EN_B and the concomitant classification of the resulting binary combinations into Yeh's version of a Grimm bond-type triangle in which there are no discrete or discontinuous classes of bonds, but rather a continuous progression of intermediate cases connecting the idealized extremes of ionic, covalent and metallic located at each of the three vertices:

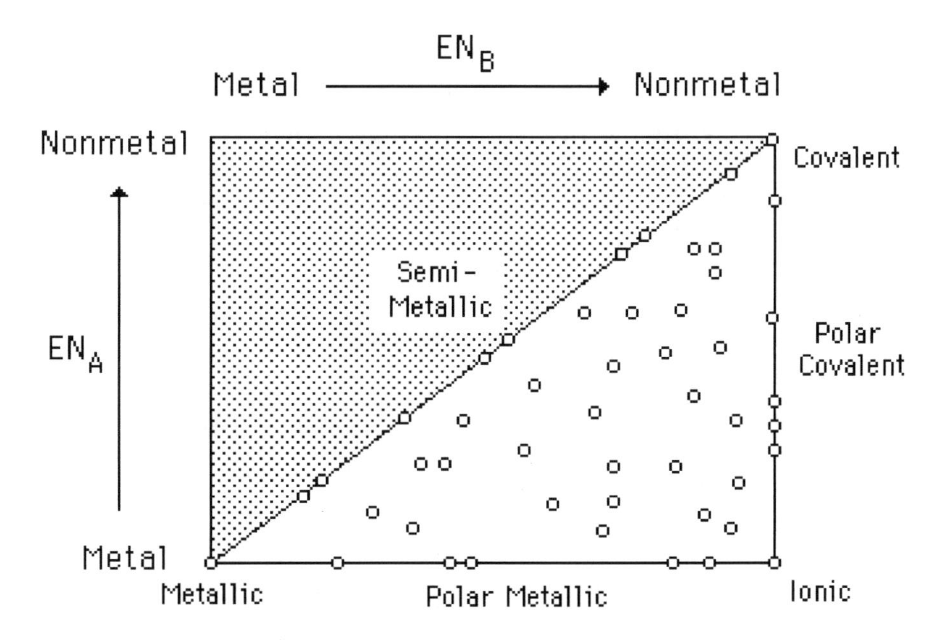

The most obvious strengths of the Fernelius triangle are the greater symmetry of its overall shape and the fact that its more complex coordinates can be crudely rationalized using simple Hückel theory – features which make it an attractive candidate for use in more advanced inorganic chemistry courses. According to simple Hückel theory, the change in energy (ΔE) upon forming a bond between two atoms (r and s) of differing electronegativity is given by the expression:

$$\Delta E = [(\Delta\alpha_{rs})^2 + 4\beta_{rs}^{\;2}]^{0.5} / 2 \qquad [4]$$

where $\Delta\alpha_{rs}$ is the difference in the coulomb integrals (α_r and α_s) for the two atoms:

$$\textit{coulomb integral} = H_{rr} = \alpha_r \qquad [5]$$

and β_{rs} is the resonance integral for the resulting bond:

$$\textit{resonance integral} = H_{rs} = \beta_{rs} \qquad [6]$$

Obviously, in the case where the two atoms are identical, $\Delta\alpha_{rs} = 0$ and equation 4 reduces to the well-known textbook result:

$$\Delta E = \beta_{rs} \qquad [7]$$

As Coulson *et al* pointed out many years ago (30):

... α_r is a measure of the electron attracting power of atom r and can be considered, in a crude way, to represent the electronegativity of that atom.

whence it follows that:

$$\Delta\alpha_{rs} \simeq \Delta EN_{rs} = \Delta EN_{AB} \qquad [8]$$

Likewise (30):

... β_{rs} can be interpreted as the electron-attracting power of the bond r-s as distinct from the individual atoms r and s which are involved in it.

and can, via the well-known Wolfsberg-Helmholtz approximation for this integral and approximation 8, be equated with the average of the electronegativities of the two bonded atoms (31):

$$\beta_{rs} = kS_{rs}(\alpha_r + \alpha_s)/2 \simeq b(EN_r + EN_s)/2 = b(EN_A + EN_B)/2 \qquad [9]$$

Thus the coordinates of the quantified Fernelius triangle may be alternatively viewed as a crude approximation of the following Hückel triangle (1):

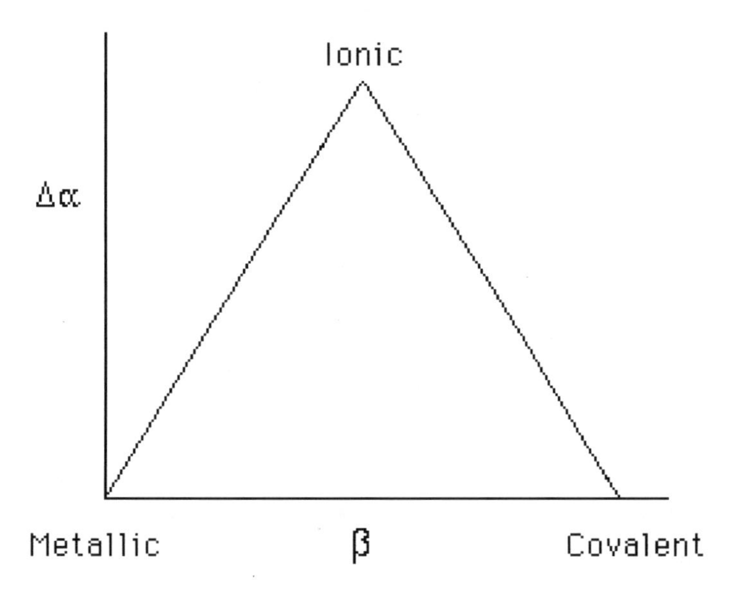

Despite its simplicity, I have never encountered a textbook, other than that of Alcock, which made use of any of the various versions of the Grimm right triangle. On the other hand, over the past 45 years a fair number of both introductory and advanced textbooks have made use of various qualitative versions of the Fernelius isosceles triangle (32-39) and, more recently, of its quantified version as well (40-43).

2.4 References and Notes

1. This lecture is based, in part, on W. B. Jensen, "The Historical Development of the Van Arkel Bond-Type Triangle," *Bull. Hist. Chem*, **1992-1993**, *13-14*, 47-59; and W. B. Jensen, "Bond-Type Triangles: An Overview," Invited paper delivered at the 82nd Canadian Society of Chemistry Conference in Toronto, 30 May - 02 June 1999.

2. H. G. Grimm, "Allgemeines über die verschiedenen Bindungsarten," *Z. Elektrochem.*, **1928**, *34*, 430-437.

3. H. G. Grimm and H. Wolff, "Über die sprungweise Änderung der Eigenschaften in Reihen chemischer Verbindungen," in P. Debye, Ed., *Probleme der modernen Physik*, Hirzel: Leipzig, 1929, pp. 172-182.

4. H. G. Grimm, "Zur Systematik der chemischen Verbindungen von Standpunkt der Atomforschung, zugleich über einige Aufgaben der Experimentalchemie," *Naturwiss.*, **1929**, *17*, 535-540, 557-564.

5. H. G. Grimm, "Das periodische System der chemischen Verbindungen vom Typ A_mB_n," *Angew. Chem.*, **1934**, *47*, 53-48.

6. H. G. Grimm, ""Die energetischen Verhältnisse im periodischen System der chemischen Verbindungen vom Typ A_mB_n," *Angew. Chem.*, **1934**, *47*, 593-561.

7. H. G. Grimm, "Wesen und Bedeutung der chemischen Bindung," *Angew. Chem.*, **1940**, *53*, 288-292.

8. W. B. Jensen, "Electronegativity from Avogadro to Pauling. Part I. Origins of the Electronegativity Concept," *J. Chem. Educ.*, **1996**, *73*, 11-20.

9. W. B. Jensen, "Electronegativity from Avogadro to Pauling. Part II. Late 19th- and Early 20th-Century Developments," *J. Chem. Educ.*, **2003**, *80*, 279-287.

10. C. W. Stillwell, "Crystal Chemistry: I. A. Graphic Classification of Binary Systems," *J. Chem. Educ.*, **1936**, *13*, 415-419.

11. C. W. Stillwell, *Crystal Chemistry*, McGraw-Hill: New York, NY, 1938, Chapter 5 and foldout chart in appendix.

12. P. Y. Yeh, "A Chart of Chemical Compounds Based on Electronegativities," *J. Chem. Educ.*, **1956**, *33*, 134.

13. Compare L. Pauling, *The Nature of the Chemical Bond*, Cornell University Press: Ithaca, NY, 1939, p. 54 and *Ibid.*, 3rd ed., 1960, p. 93.

14. N. W. Alcock, *Bonding and Structure: Structural Principles in Inorganic and Organic Chemistry*, Horwood: Chichester, 1990, pp. 18-24.

15. Many of these alternative orientations were later reported by Sproul without acknowledgement of the earlier literature. See references 28 and 29.

16. W. C. Fernelius, R. F. Robey, "The Nature of the Metallic State," *J. Chem. Educ.*, **1935**, *12*, 53-68.

17. A. E. van Arkel, *Moleculen en Kristallen*, van Stockum & Zoon: Gravenhage, 1941. The second Dutch edition was translated as A. E. van Arkel, *Molecules and Crystals*, Butterworths: London, 1949. The triangle appears on page 205 of this translation. I have assumed, but not personally verified, that the triangle also appears in the first Dutch edition.

18. J. A. A. Ketelaar, *Chemical Constitution*, 2nd ed., Elsevier: Amsterdam, 1958, p. 21. Once again, I have assumed, but not personally verified, that the triangle also appears in the first Dutch edition of 1947. Ketelaar overlapped with van Arkel at Leyden in the period 1934-1941, during which time he collaborated with van Arkel's group on X-ray crystallography. Hence, it is likely that their diagrams are related developments.

19. W. L. Jolly, *Modern Inorganic Chemistry*, McGraw-Hill: New York, NY, 1984, p. 285.

20. L. C. Allen, "Extension and Completion of the Periodic Table," *J. Am. Chem. Soc.*, **1992**, *114*, 1510-1511. In fact Allen's brief note provides virtually no explanation of his diagram, and the one given here is based on personal discussions held with Allen at Princeton University in July of 1992.

21. Allen originally used the term "spectroscopic electronegativities" to describe his scale, as they were based on the averaging of the spectroscopic orbital energies of an atom's valence electrons. When more and more critics pointed out that many of the unique predictions claimed by Allen for his scale were merely scale translations of already well-known electronegativity correlations, Allen attempted to dissociate his scale from the electronegativity literature by renaming his values "valence-shell energies" instead and using their absolute values rather than their Pauling adjusted values. More recently he has renamed his scale once more and now prefers the label "configuration energies."

22. W. B. Jensen, "Quantity or Quality?," *Educ. Chem.*, **1994**, *31*, 10.

23. W. B. Jensen, "A Quantitative van Arkel Diagram," *J. Chem. Educ.*, **1995**, *72*, 395-398.

24. The author's version was actually first presented at a departmental "Symposium on Chemical Bonding" held at the University of Wisconsin in July of 1980 and was used for many years in his teaching before finally being published.

25. L. C. Allen, J. F. Capitaini, G. A. Kolks, G. D. Sproul, "Van Arkel - Ketelaar Triangles," *J. Mol. Struct.*, **1993**, *300*, 647-655.

26. G. D. Sproul, "Electronegativity and Bond Type: 1. Tripartite Separation," *J. Chem. Educ.*, **1993**, *70*, 531-534.

27. G. D. Sproul, "Electronegativity and Bond Type: 2. Evaluation of Electro-

negativity Scales," *J. Chem. Phys.*, **1994**, *98*, 6699-6703.

28. G. D. Sproul, ""Electronegativity and Bond Type: 3. Origins of Bond Type," *J. Chem. Phys.*, **1994**, *98*, 13221-13224.

29. G. D. Sproul, "Electronegativity and Bond Type: Predicting Bond Types," *J. Chem. Educ.*, **2001**, *78*, 387-390.

30. C. A. Coulson, B. O'Leary, R. B. Maillon, *Hückel Theory for Organic Chemistry*, Academic Press: London, 1979, p. 27.

31. D. B. Cook, *Structures and Approximations for Electrons in Molecules*, Harwood: Chichester, 1978, p. 171.

32. K. B. Harvey, G. B. Porter, *Introduction to Physical Inorganic Chemistry*, Addison-Wesley: Reading, MA, 1963, pp. 1 and 4.

33. CBA, Chemical Bond Approach, *Chemical Systems*, McGraw-Hill: New York, NY, 1965, p. 594.

34. M. B. Ormerod, *The Architecture and Properties of Matter: An Approach through Models*, Arnold: London, 1970, p. 103.

35. D. M. Adams, *Inorganic Solids*, Wiley: New York, NY. 1974, p. 106.

36. K. M. Mackay, R. A. Mackay, *Introduction to Modern Inorganic Chemistry*, 4th ed., Blackie: Glasgow, 1989, p. 80.

37. J. D. Lee, *Concise Inorganic Chemistry*, 4th ed., Chapman & Hall: London, 1991, p. 31.

38. M. S. Silberberg, *Principles of General Chemistry*, 2nd ed., McGraw-Hill, Boston, MA, 2010, p. 281.

39. J. Gilman, *Electronic Basis of the Strength of Materials*, Cambridge University Press; Cambridge, 2003, p. 67.

40. G. M. Bodner, L. H. Richard, J. N. Spencer, *Chemistry: Structure and Dynamics*, Wiley: New York, NY, 1996, pp. 186-189, 352.

41. N. C. Norman, *Periodicity and the s- and p-Block Elements*, Oxford University Press: Oxford, 1997, pp, 56-58 and cover.

42. G. Wulfsberg, *Inorganic Chemistry*, University Science Books: Sausalito, CA, 2000, pp. 775-778.

43. G. Raynor-Canham, T. Overton, *Descriptive Inorganic Chemistry*, Freeman, New York, NY, 2006, pp. 102-103.

III
Interconversions

3.1 The Choice of Electronegativity Scale

Before discussing the mathematical interrelationship between the Grimm and Fernelius versions of the bond-type triangle, it is necessary to say something about the choice of electronegativity (*EN*) scale used to construct the various coordinate systems (1). It has been estimated that more than 25 different *EN* scales have been proposed in the chemical literature since 1932 (2). Since, with respect to the values assigned to the main-block elements, these various scales generally display linear interscale correlation coefficients of 0.95 or better with one another, it makes little difference, from an empirical point of view, which scale is used to construct a bond-type triangle. Though specific points may shift slightly, the overall pattern will remain basically invariant. About the only empirical requirement is that the scale in question be complete, including values for the noble gases, since it considered bad form to intermix values from different scales on the same plot.

When it comes to interpreting the significance of the resulting diagrams, however, the choice of *EN* scale is far more important, since, as we will see in greater detail in the next lecture, these scales may be divided into two large categories, known as primary and secondary *EN* scales, respectively, and only those belonging to the first of these groups are amendable to simple interpretation. For the present, however, these considerations need not concern us and we will focus instead on optimizing the empirical aspects of the diagrams in order to facilitate an understanding of their mutual interrelationships.

3.2 Rescaling and Fractional Electronegativities

The following two schematic plots compare the coordinates of the three vertices of Stillwell's orientation of the Grimm triangle with those of the vertices of the Fernelius triangle as calculated using the well-known Pauling "thermochemical" *EN* scale ($EN_{Cs} = 0.79, EN_F = 3.98$):

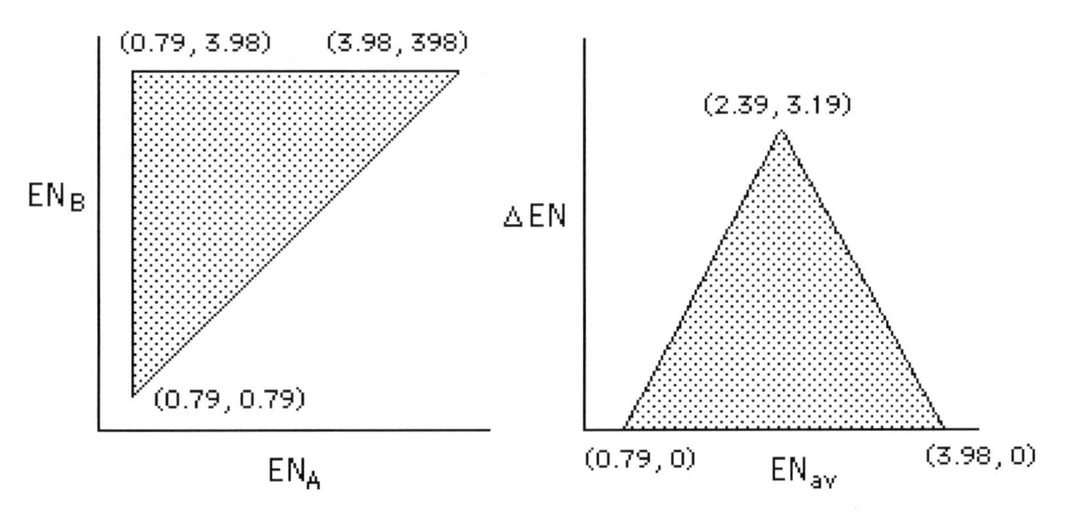

Two problems are immediately apparent. First, neither plot is positioned on the origin of its coordinate system, and, second, the numerical values of the vertex coordinates will change as the choice of *EN* scale changes, even though the relative distribution pattern of various AB points within the triangles will remain relatively invariant. Thus, if one employs the Sanderson "stability ratio" *EN* scale instead of the Pauling scale, the vertex coordinates for the Grimm triangle shift from (0.79, 0.79), (3.98, 3.98), and (0.79, 3.98) to (0.49, 049), (5.75, 5.75) and (0.49, 5.75) respectively.

Pauling originally assigned H an arbitrary *EN* value of 2.00 and then determined the electronegativities of the other elements relative to that choice, giving a scale that varied between roughly 0.5 and 4.00 (3). Most alternative *EN* definitions give a very different numerical range for their resulting *EN* values but have been artificially adjusted using linear correlation equations so as to conform as close as possible with Pauling's original arbitrary choice.

Thus, for example, if one were to use the unadjusted or absolute *EN* values calculated by means of either the Mulliken or Allred-Rochow *EN* definitions, the resulting vertex shifts would be even greater than those illustrated above for the Pauling adjusted Sanderson scale.

All of these irritating, albeit ultimately trivial, problems may be easily eliminated by adopting both a common numerical origin and a common numerical terminal point for all current *EN* scales, and rescaling the individual *EN* values for the remaining elements as fractions of the distance separating these two common reference points. Adopting the most chemically active electropositive element (Cs) at the origin and the most chemically active electronegative atom (F) as the terminal point, one may easily calculate the corresponding *fractional electronegativity* or *FEN* value for any element *X*, irrespective of the particular *EN* definition, using the following equation:

$$FEN_X = (EN_X - EN_{Cs})/(EN_F - EN_{Cs}) \quad [1]$$

Whatever the definition of choice, this equation automatically sets the *FEN* value of Cs at 0.00 and that of F at 1.00. Since Fr is generally assigned an *EN* value equal to or slightly greater (due in part to relativistic effects) than that of Cs, there are no negative *FEN* values and only two *FEN* values greater than 1.00 (He and Ne), both of which correspond to chemically inactive elements not normally plotted on bond-type triangles. A selection of typical *FEN* values for both the H-He and the main-block elements, based on the Bohr electronegativity definition described in Appendix I, is shown in the table on the following page. It should also be noted that the resulting *FEN* values are by definition unitless, thereby obviating the problem of variable units which usually plagues the direct comparison of the various *EN* scales. Whereas a direct numerical comparison, for example, of the *EN* values for C as calculated using the Bohr (1.57), Allred-Rochow (2.50), Pauling (2.55), and Sanderson (3.79) scales is invalid because each of these scales has a

H	He
0.50	1.02

Li 0.05	Be 0.24		B 0.38	C 0.54	N 0.70	O 0.85	F 1.00	Ne 1.16
Na 0.04	Mg 0.19		Al 0.29	Si 0.40	P 0.52	S 0.62	Cl 0.74	Ar 0.85
K 0.02	Ca 0.13	Zn 0.24	Ga 0.31	Ge 0.40	As 0.50	Se 0.60	Br 0.69	Kr 0.78
Rb 0.01	Sr 0.12	Cd 0.22	In 0.28	Sn 0.38	Sb 0.46	Te 0.55	I 0.62	Xe 0.71
Cs 0.00	Ba 0.10	Hg 0.25	Tl 0.30	Pb 0.38	Bi 0.46	Po 0.54	At 0.62	Rn 0.70
Fr 0.01	Ra 0.11	112	113	114	115	116	117	118

different origin and a different numerical span, the same is not true of a direct comparison of their corresponding rescaled *FEN* values as calculated using equation 1. This gives the results: Allred-Rochow (0.50), Bohr (0.54), Pauling (0.55), and Sanderson (0.63), thus showing that all four definitions place this element between roughly half and two thirds of the distance separating the extremes of Cs and F. Note that this rescaling does not effect the relative order of the *EN* values on a given scale nor their relative distances of separation. Thus the linear correlation coefficient for a plot of the *FEN* values of one scale versus those of another is identical to that obtained for a direct plot of their corresponding *EN* values.

Nor does the use of *FEN* values, rather than direct *EN* values, alter in any way the distribution of AB points within either the Grimm or Fernelius versions of the bond-type triangle, as may be illustrated by comparing the version of the quantified Fernelius triangle shown at the top of the following page, which is based on the direct use of Bohr *EN* values (and shows the

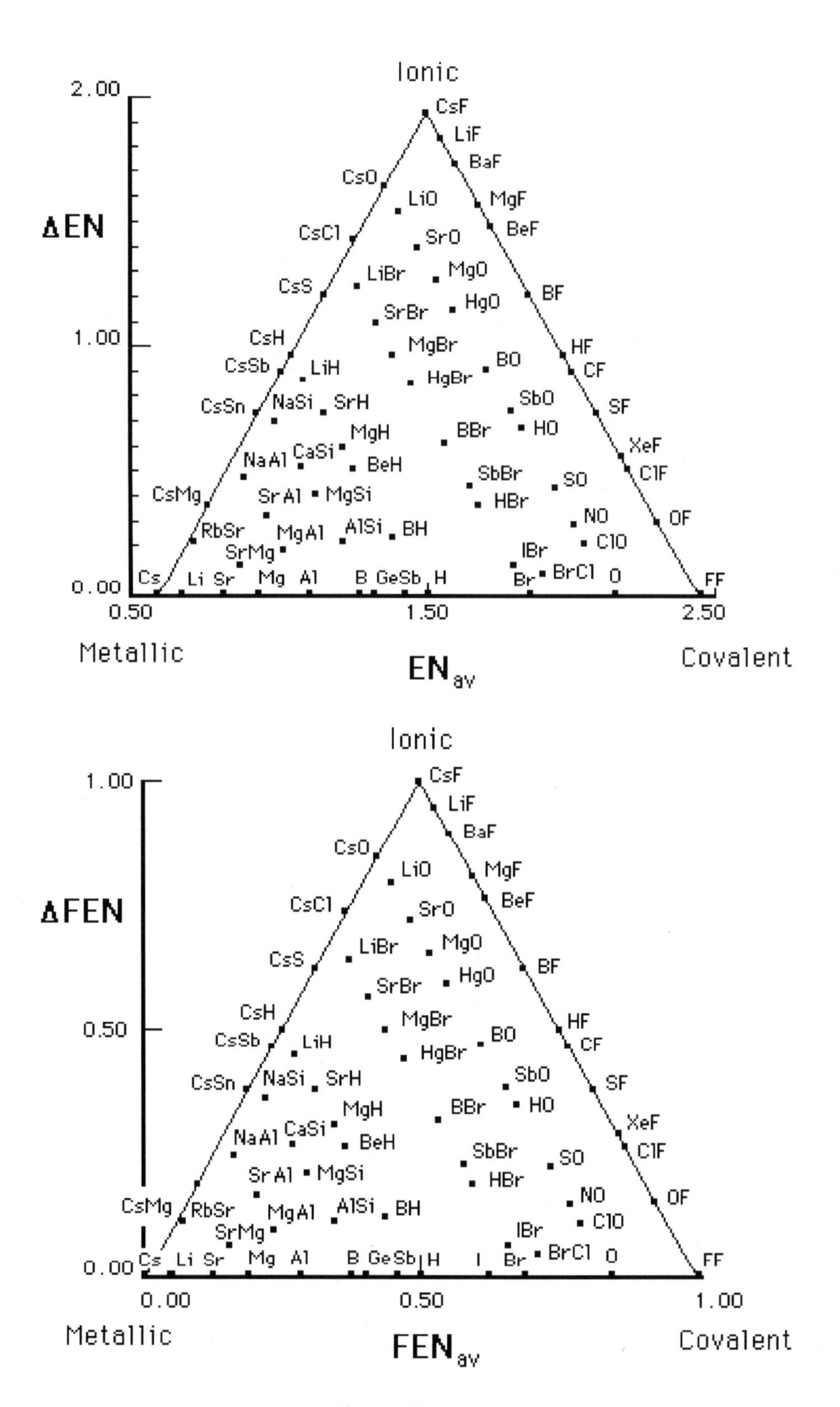
Ionic
2.00
1.00
0.00
ΔEN
CsF
LiF
BaF
MgF
BeF
BF
HF
CF
SF
XeF
ClF
OF
FF
CsO
CsCl
CsS
CsH
CsSb
CsSn
CsMg
RbSr
Cs
LiO
SrO
LiBr
MgO
HgO
SrBr
MgBr
HgBr
BO
LiH
NaSi
SrH
SbO
HO
MgH
BBr
CaSi
NaAl
BeH
SbBr
SO
HBr
SrAl
MgSi
NO
MgAl
AlSi
BH
ClO
SrMg
IBr
BrCl
Li Sr Mg Al B GeSb H Br O
0.50
1.50
2.50
Metallic
EN_{av}
Covalent
Ionic
1.00
0.50
0.00
ΔFEN
I
0.00
0.50
1.00
Metallic
FEN_{av}
Covalent

location of individual AB bonds rather than example binary compounds and simple substances) with that given at the bottom of the page, which is based on use of the corresponding Bohr *FEN* values.

What this rescaling does, however, is to make both the coordinate scales and the location of the resulting bond triangles within the coordinate system invariant to the particular choice of electronegativity definition. In addition, as may be seen from the following schematic plots, it automatically positions both the Grimm and Fernelius versions of the triangle at the origin and leads to a simplification of the resulting vertex coordinates:

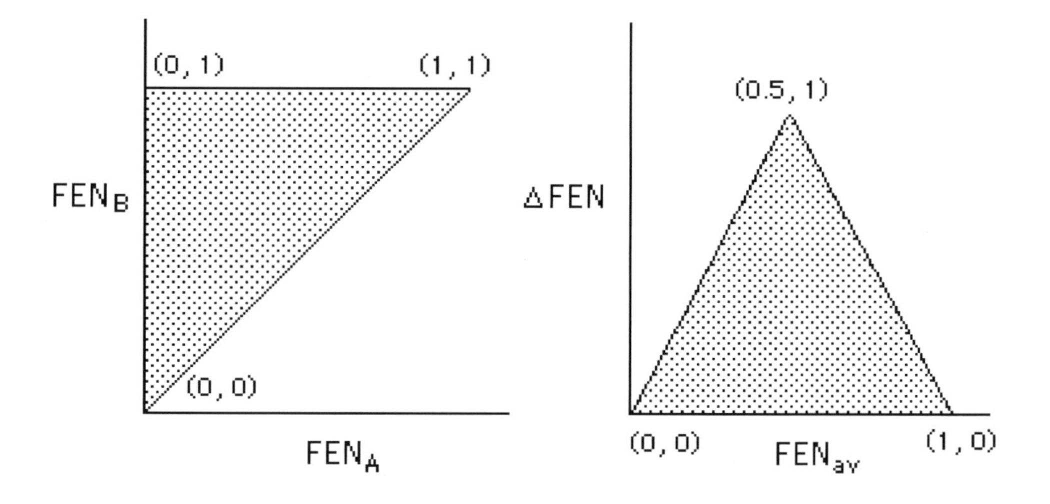

As we will discover in the next lecture, the use of *FEN* values, rather than conventional *EN* values, in our plots will also allow us to construct a more attractive quantitative version of the bond-type triangle – one which will not only eliminate the necessity of explicitly showing the x - and y-axes, but which will also lead to an even more rigorous method of characterizing individual bond types.

3.3 Coordinate Transformations

The coordinate transform which relates the Grimm and Fernelius versions of the bond-type triangle may be deduced by means of the following diagrams and is greatly facilitated by the use of the *FEN* coordinates:

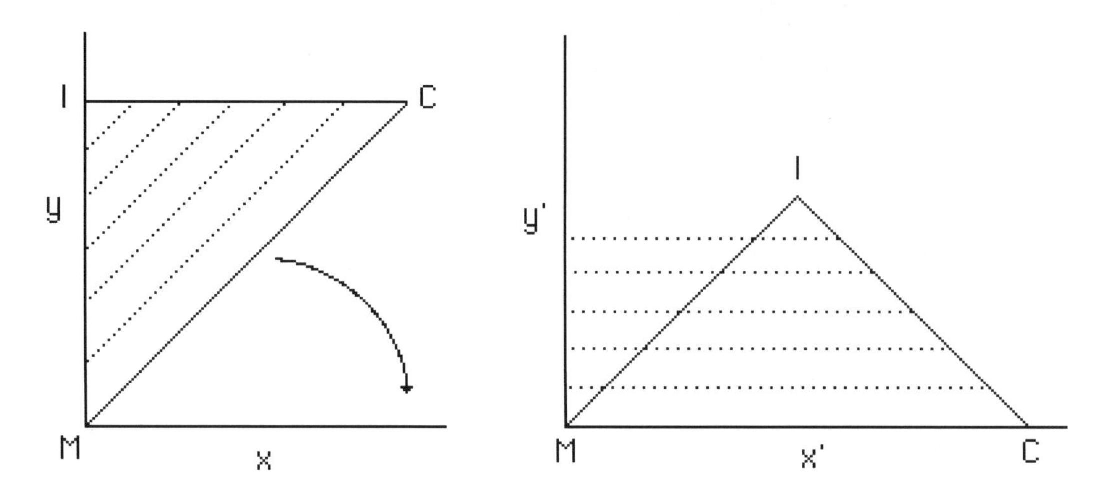

All of the dotted diagonals shown on the Stillwell orientation of the Grimm triangle on the left have a slope of +1 but variable values (b) for their intercepts with the y-axis. Thus they correspond to the simple linear equation:

$$y = x + b \qquad [2]$$

When this triangle is rotated 45°, as shown on the right, to produce a flatten version of the Fernelius triangle, these various b intercepts become the corresponding coordinates along the new y'-axis. Thus by equation 2:

$$y' = b = y - x \qquad [3]$$

or, given that $y = FEN_B$ and $x = FEN_A$:

$$y' = FEN_B - FEN_A = \Delta FEN \qquad [4]$$

as required.

Likewise, all of the dotted diagonals shown below on the left for the Grimm triangle have a slope of -1 but variable values (c) for their intercepts with the y-axis.

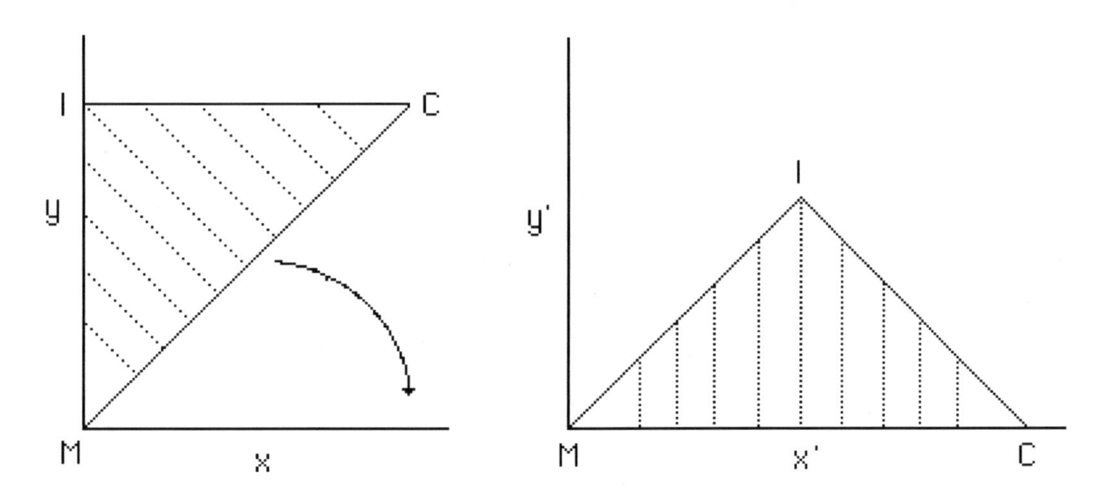

Thus they all correspond to the simple linear equation:

$$y = -x + c \qquad [5]$$

When this triangle is rotated 45°, as shown on the right, to produce a flatten version of the Fernelius triangle, these various c intercepts become the corresponding coordinates along the new x'-axis. Thus by equation 5:

$$x' = c = x + y \qquad [6]$$

or, given that $y = FEN_B$ and $x = FEN_A$:

$$x' = FEN_A + FEN_B \qquad [7]$$

As is apparent from this latter result, the version of the Fernelius triangle produced by the 45° rotation of the Grimm triangle is not quite identical to the usual version, but is more flattened due to an apex value of 90°. However, as shown in the following figure, this apex angle may arbitrarily changed by the simple act of multiplying the x' coordinate values by a fractional scaling factor, which for the technical reasons given in the final section of the previous lecture, has been chosen to be 0.5:

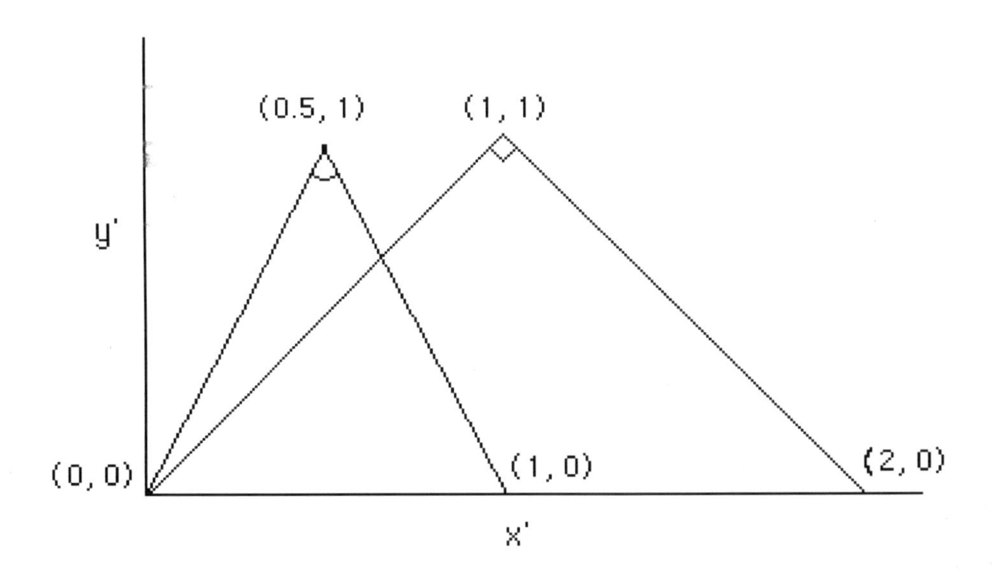

Thus:

$$x'' = 0.5x' = (FEN_A + FEN_B)/2 = FEN_{av} \quad [8]$$

as required.

3.4 Branching

Examination of the various data points in a quantified bond-type triangle quickly reveals that they form a simple pattern. All of the binary AB combinations or bonds for a given element lie on two straight lines which intersect the base of the triangle connecting the metallic and covalent vertices at the point which corresponds to the element's homonuclear bond or simple substance. The right-hand line or branch contains all of the binary combinations or bonds in which the element in question is the more electro-positive or cationic component, whereas the left-hand branch contains all of the binary combinations or bonds in which it is the more electronegative or anionic component. For example, the characteristic electropositive branch for the element hydrogen (corresponding to various H-B bonds) and the characteristic electronegative branch (corresponding to various "hydride" or A-H bonds) are shown in the following figures for both the Grimm and Fernelius versions of the bond-type triangle:

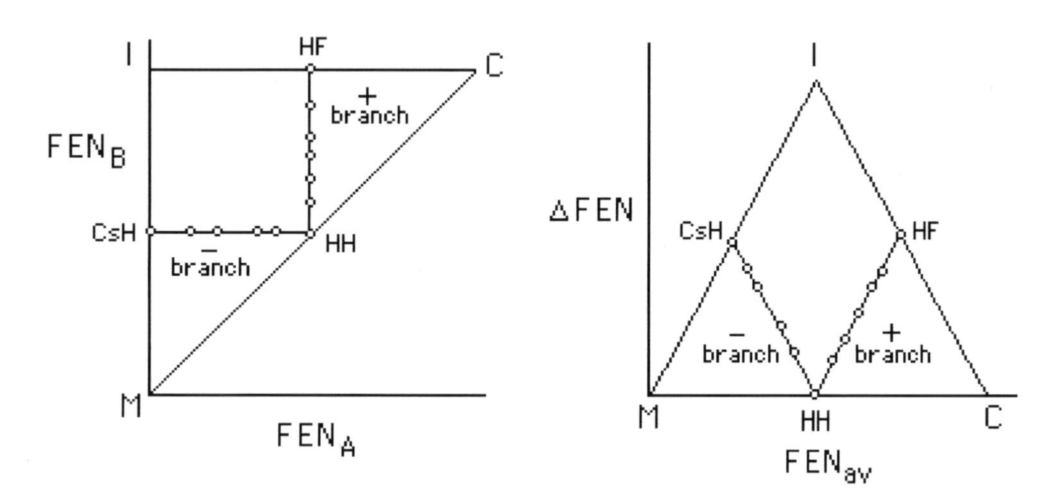

Inspection of these figures shows that the electropositive branch is always parallel to the edge of the triangle connecting the ionic and metallic vertices and intersects with the opposite edge connecting the ionic and covalent vertices at the point corresponding to the element's bond with fluorine. Likewise, the electronegative branch is always parallel to the edge of the triangle connecting the ionic and covalent vertices and intersects with the opposite edge connecting the ionic and metallic vertices at the point corresponding to the element's bond with cesium. Indeed, the metallic-ionic edge of the triangle is merely the electropositive branch for Cs, which, because it is the most electropositive chemically active element, has no corresponding electronegative branch. Likewise the covalent-ionic edge of the triangle is actually the electronegative branch for F, which, because it is the most electronegative chemically active element, has no corresponding electropositive branch (4).

Obviously the various AB points within the body of the triangle result from the intersection of the electropositive branch of one element with the electronegative branch of another, as explicitly illustrated below for the five elements, Ca, Mg, H, O, and F:

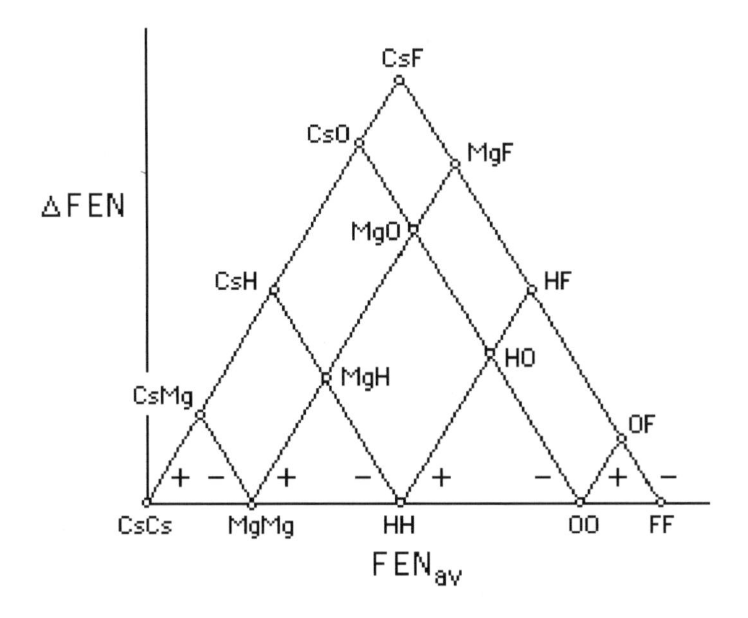

3.5 References and Notes

1. This chapter is based, in part, on the lecture: W. B. Jensen, "Bond-Type Triangles: An Overview," Invited paper delivered at the 82nd Canadian Society of Chemistry Conference in Toronto, 30 May - 02 June 1999; and on the unpublished manuscript: W. B. Jensen, "Phase Diagrams and Bond Triangles," University of Cincinnati, November 2003.

2. Estimated by author. However, as repeatedly noted by Allen, the vast majority of textbooks employ either the Pauling or Allred-Rockow scales.

3. The value for H on the original Pauling scale has since been adjusted from the original value of 2.00 to the value 2.20.

4. W. B. Jensen, "A Quantified van Arkel Bond-Type Triangle," *J. Chem. Educ.*, **1995**, *72*, 395-398.

IV

Prediction or Definition?

4.1 Misconceptions and Circular Reasoning

As we saw in Lecture II, as early as 1956 Yeh had divided his version of the quantified Grimm triangle into three discrete regions, corresponding to ionic, covalent and metallic compounds (1):

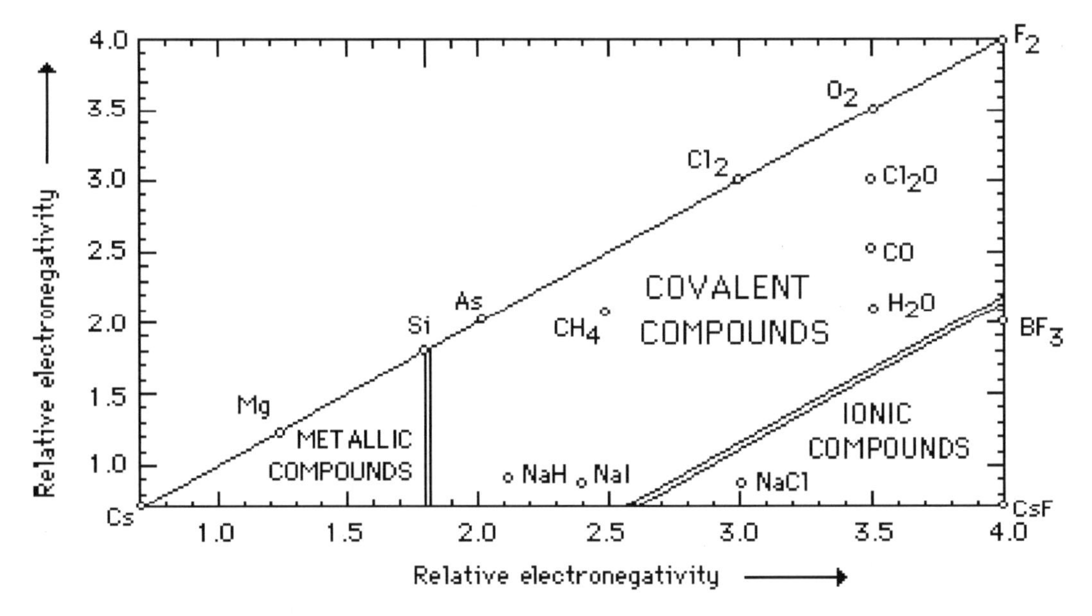

and, more recently, Sproul has done the same with his quantified version of the triangle, though he disagrees with Yeh with regard to the orientation of the boundary separating the covalent and ionic regions (2). Sproul calls this division of the triangle into discrete bonding regions with sharp boundaries the "separation postulate," and his particular choice for these boundaries is illustrated for both the Grimm and Fernelius versions of the triangle in the following schematic diagrams using *FEN* coordinates rather than the corresponding *EN* coordinates (3):

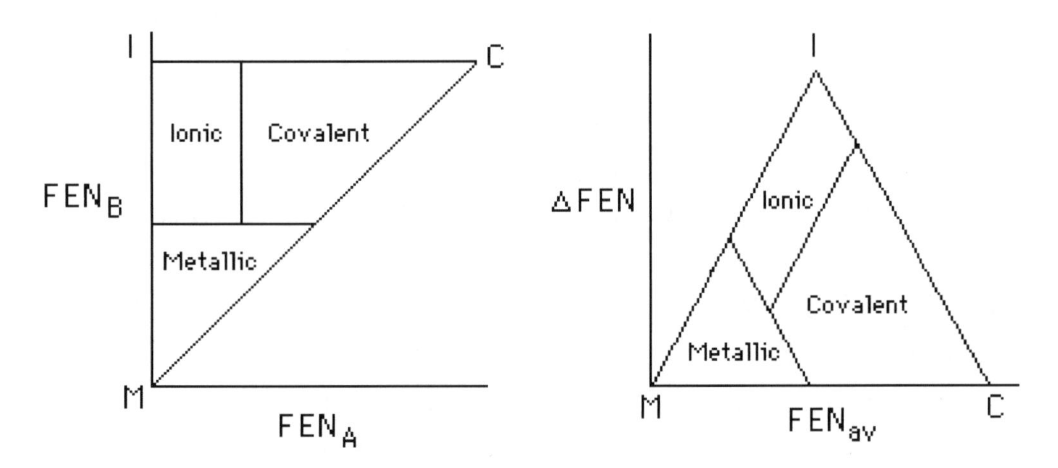

Likewise, Allen has also imposed discrete boundaries on the Fernelius bond triangle as shown below (where their equivalent appearance on the Grimm triangle is also given for comparison), but has advocated a fourfold, rather than a threefold, division, corresponding to ionic, covalent, metallic, and so-called metalloid-band (MB) compounds (4):

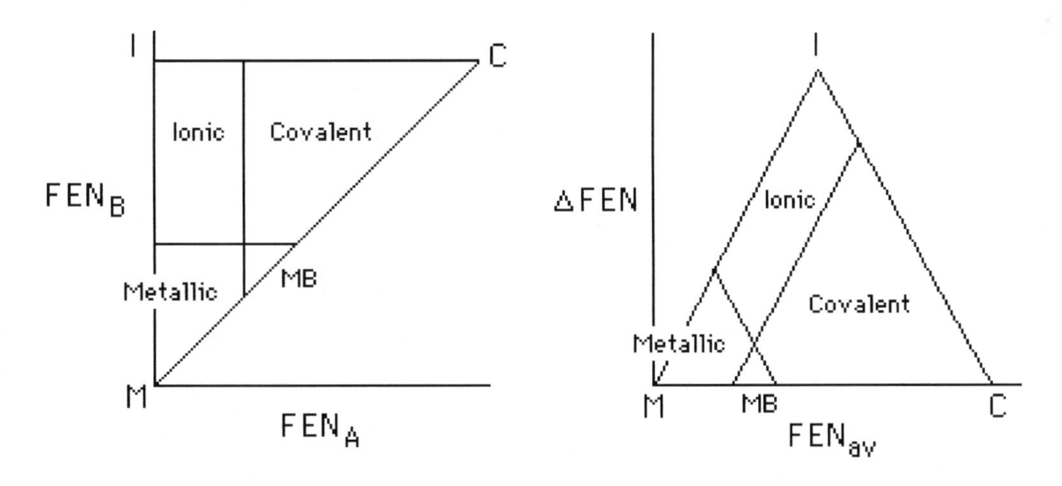

There are at least three serious problems with imposing boundaries of this sort on a quantified bond-type diagram. The first and most important is that such a practice subverts the entire purpose of these diagrams, which was and is to explicitly illustrate the absence of discrete either/or boundaries separating ionic, covalent and metallic bonds and to show instead that, in reality, these three extremum idealizations gradually and progressively shade into one another through a series of intermediate cases which partake to varying degrees of the characteristics of the ideal cases located at the vertices.

Secondly, even if one is willing to accept the so-called separation postulate, none of the above authors has provided independent empirical and/or theoretical criteria for determining where the various boundaries should be drawn. Thus Yeh gave no rationale at all for his particular choice of boundaries, whereas Sproul based his choice on the optimal sorting of 311 binary compounds and an unspecified number of simple substances which had been characterized as being primarily ionic, covalent or metallic by A. F. Wells in his well-known reference work on the structures of inorganic compounds (2, 4). However, consultation of this book reveals that Wells provided no independent quantitative and/or theoretical criteria for these characterizations other than a qualitative, and largely intuitive, consideration of the relative positions the component atoms in the periodic table, which is, of course, equivalent to a qualitative consideration of their relative *EN* values. In other words, Sproul's boundaries are based on circular reasoning. All he has shown is that a quantitative bond-type diagram which is explicitly based on electronegativity does a relatively good job of reproducing an intuitive classification of binary compounds which is implicitly based on electronegativity. Though it helps us to visualize and quantify Wells' personal guesses, it provides no independent physical or quantitative justification for their accuracy or objectivity. This conclusion also calls into question Sproul's further claims that the resulting boundaries based on this essentially "soft" qualitative data set can be used not only to evaluate the relative accuracy of various competing electronegativity scales (5) but also to dispel the "myth" that ionic character depends on electronegativity differences alone (6, 7).

The same is equally true of the boundaries in Allen's four-region diagram. These are based Allen's claim that his spectroscopic *EN* scale is uniquely able to predict the so-called metalloid band in the periodic table (8). Having determined which elements are metalloids as a function of their *EN* values allows Allen to determine the location of the small triangular metalloid band (MB) region shown on his version of the bond triangle. He simply locates the positions of the most electronegative metalloid (As) and the most electropositive metalloid (Si) along the base connecting the metallic and covalent extremes and draws the lines for the electropositive branch for the various Si bonds and the electronegative branch for the various As bonds. Their intersection (SiAs) determines the apex of the resulting metalloid triangle, and their further extension until they intersect the remaining sides of the complete triangle automatically determines both the ionic-metallic boundary and the ionic-covalent boundary. As is apparent in the following detailed diagram, because of the small range of *EN* values assigned to the metalloids by Allen, the resulting MB triangle is actually quite small:

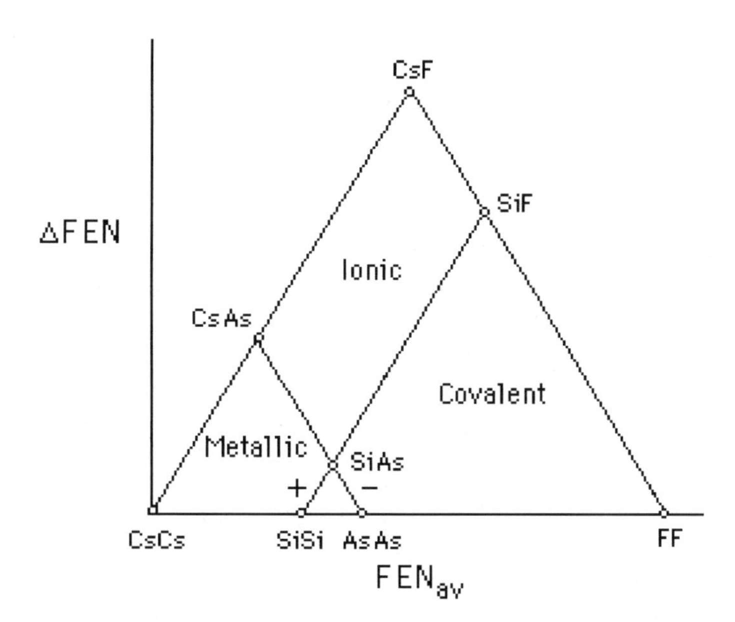

As with Sproul, examination of Allen's assumptions quickly reveals that his boundaries are also based on circular reasoning since his original prediction of the so-called metalloid band in the periodic table is not based on the verification of independent quantitative and/or theoretical criteria for which elements are or are not metalloids. He merely uses the definition given by Rochow in his small monograph on the metalloids – a definition which was, in turn, explicitly based on their relative electronegativity values (9):

In between the metals and nonmetals there is a collection of elements of electronegativity 1.8 to 2.2 which are neither metals nor nonmetals, and which are best described as metalloids.

Since Allen's spectroscopic *EN* scale shows a 0.98 linear correlation coefficient with the Allred-Rochow electrostatic *EN* scale used by Rochow to define the metalloid band in the first place, it is hardly surprising that Allen's scale also reproduces the same band. Thus Allen's boundaries merely show that a diagram based explicitly on *EN* values can reproduce a definition of metalloids which was, in turn, also explicitly based on *EN* values.

Thirdly, Allen makes no attempt to verify, either by means of theory or empirical evidence – such as the sorting of actual compounds (as done by Sproul), band gaps, or electrical conductivities – his further assumption that simple extensions of the edges of the metalloid triangle automatically determine the ionic-covalent and ionic-metallic boundaries as well. The fact that the ionic-covalent boundary is identical to the electropositive branch for Si means that all bonds lying above the Si-As bond in which Si is the more electropositive partner, be they Si-H or Si-F, are equally borderline ionic in character and, similarly, the claim that the ionic-metallic boundary is identical to the electronegative branch for As means that all bonds lying above Si-As bond in which As is the more electronegative partner, be they Cs-As or B-As, are equally borderline metallic in character – claims which even the crudest level of chemical intuition would consider inherently improbable.

Close examination of Sproul's diagram shows that, though his ionic-covalent and ionic-metallic boundaries do not coincide with those of Allen, they are, nonetheless, also drawn parallel to the opposite edges of the triangle.

Thus they also parallel the electropositive and electronegative branches for the bonds of various elements and are subject to exactly the same criticisms (10).

Regrettably, the majority of textbooks that have cited the quantitative Fernelius triangle have also reproduced either Sproul's threefold or Allen fourfold division of compounds and simple substances, along with all of the above dubious implications (11-13). Indeed, in some case, these have been further exacerbated by additional errors on the part of the textbook authors themselves (14).

4.2 Triangular Coordinates

Having rejected the so-called separation postulate for lack of both a theoretical and an independent empirical basis, the question now becomes one of how to make the alternative continuum interpretation of the bond-triangle more pedagogically accessible. As shown in the diagram on the below, the key

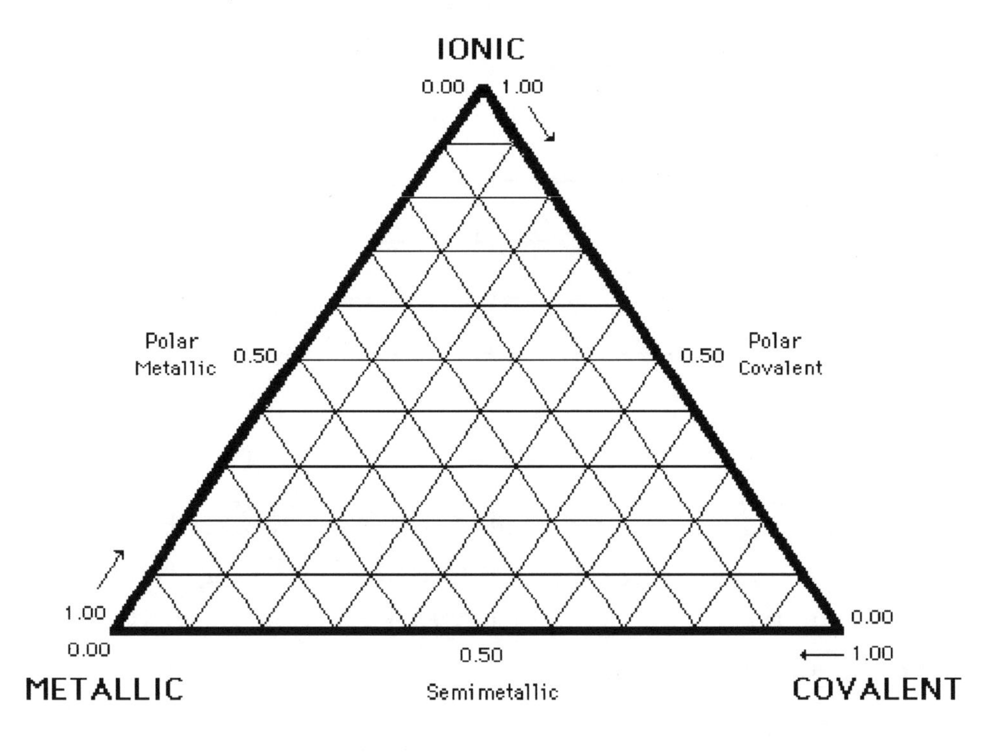

to this task lies in exploiting the advantages of our rescaled fractional electronegativity or *FEN* coordinates to the fullest by imposing a grid of triangular coordinates on the resulting Fernelius version of the bond triangle similar to those used for such plots as ternary phase diagrams, Maxwell color triangles, and ternary glass composition triangles (15-18).

As shown in the following three diagrams, the triangular coordinates are created by simultaneously superposing three separate grid systems on the bond triangle. The first of these measures the fractional distance from the ionic vertex to the opposite edge using a series of lines parallel to the edge in question, the second does the same for the covalent vertex, and the third does the same for the metallic vertex:

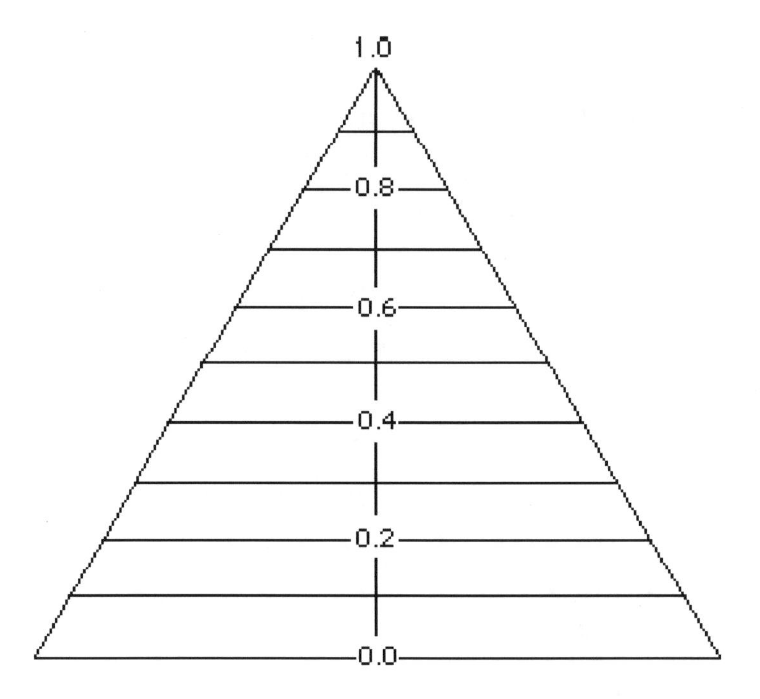

In each case the grid is assigned a value of 1.00 at the vertex in question and a value of 0.00 at the opposite edge, the intermediate locations having fractional values which increase as one approaches the vertex and decrease as one approaches the opposite edge.

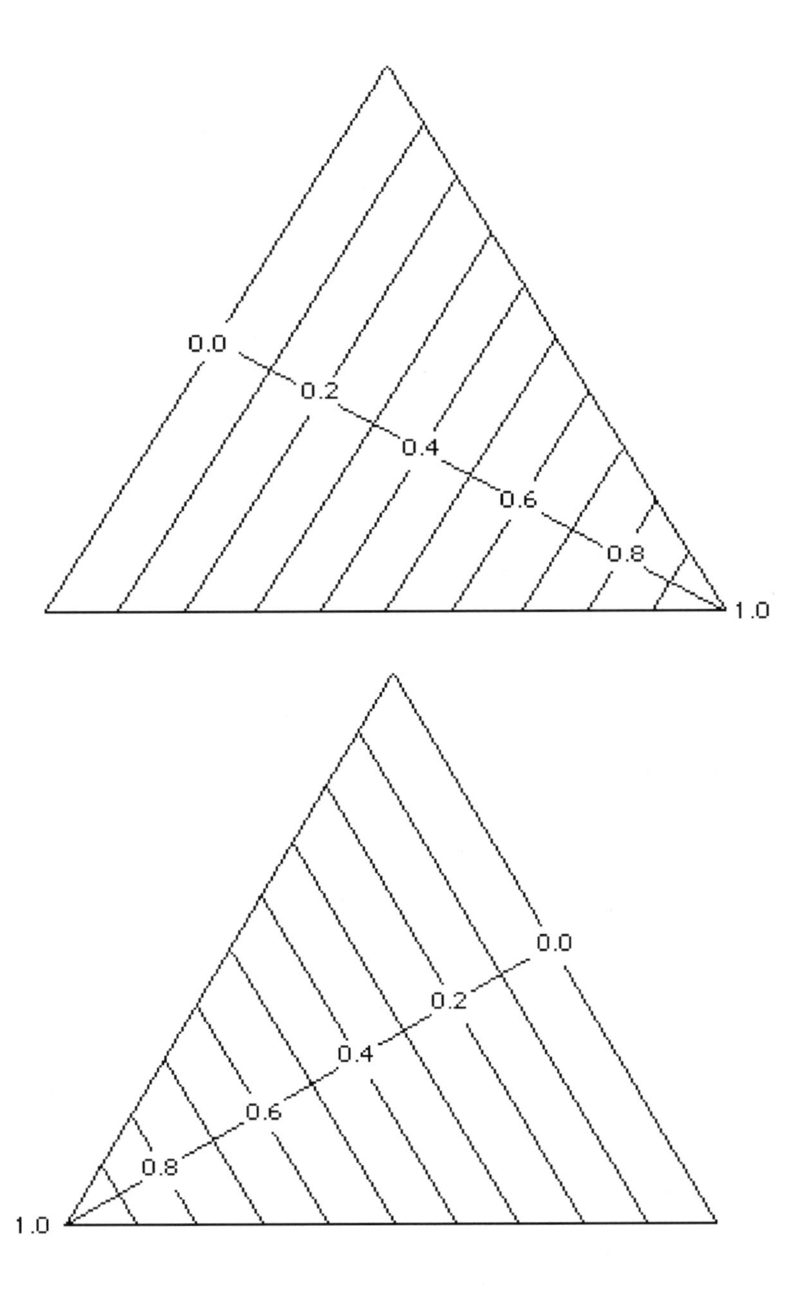

One still plots the points in the bond triangle using the rectilinear ΔFEN and FEN_{av} coordinates, but then recharacterizes the resulting location of each data point using its corresponding triangular coordinates – in other words, using the three fractions which express its relative distances from each of the three idealized vertices. The fractional distance from the ionic vertex defines the fractional ionicity (f_i) of the bond or point in question, its fractional distance from the covalent vertex defines its fractional covalency (f_c), and its fractional distance from the metallic vertex defines its fractional metallicity (f_m). From the mathematics of triangular coordinate systems, the values of these three numbers for any given bond or point must automatically sum to 1.00, just as the mole fractions do in a ternary phase diagram:

$$f_i + f_c + f_m = 1.00 \qquad [1]$$

4.3 Fractional Bond Character

Assignment of these fractional coordinates may be done by making transparent overlays of each of the three grids shown above and laying them on the bond triangle in order to successively read off the three fractional values, but this is both awkward and imprecise. Rather it far simpler to calculate the fractional bond characters of each point using the *FEN* values of the component atoms. Inspection of the grid for fractional bond ionicity shows that it is identical to the rectilinear *y*-coordinate for the Fernelius triangle:

$$f_i = \Delta FEN = FEN_B - FEN_A \qquad [2]$$

Likewise the grid system for fractional bond covalency parallels the electropositive branches of the simple substances plotted along the base of the triangle and is therefore numerically identical to the *FEN* value of the more electropositive bond component:

$$f_c = FEN_A \qquad [3]$$

Lastly, the grid system for fractional bond metallicity parallels the electronegative branches of the simple substances plotted along the base of the triangle. However, since it decreases rather than increases as the *FEN* value of the more electronegative bond component increases, it is given by its complement:

$$f_m = 1 - FEN_B \qquad [4]$$

As may be readily verified, in keeping with equation 1, these three equations sum to 1.00.

The validity of these equations is even more apparent when they are applied to the Grimm version of the bond triangle. Here the grid for the fractional covalency directly coincides with the *x*-coordinate or FEN_A and the grid for the fractional metallicity directly coincides with the compliment, $1 - FEN_B$, of the *y*-coordinate or FEN_B:

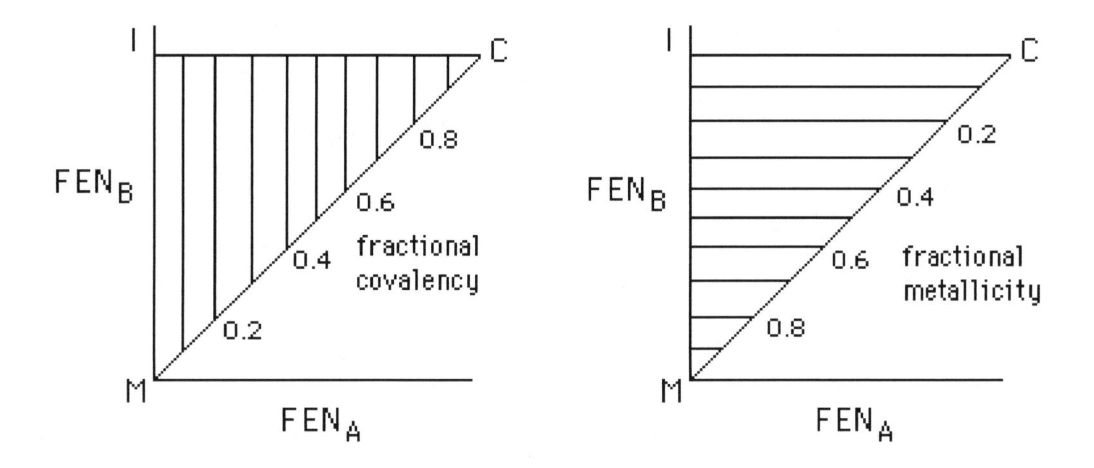

Likewise, as shown in the diagram at the top of the next page, the fractional ionicity coincides with the positive diagonals which, as we demonstrated in the previous lecture, correspond to ΔFEN:

It is of interest to note that all systems based on the use of triangular coordinates display the same underlying mathematical (but not physical) isomorphism (15). Thus not only bond triangles, but also ternary phase diagrams, Maxwell color triangles, and ternary glass composition diagrams may all be alternatively represented using right triangles as well as isosceles and/or

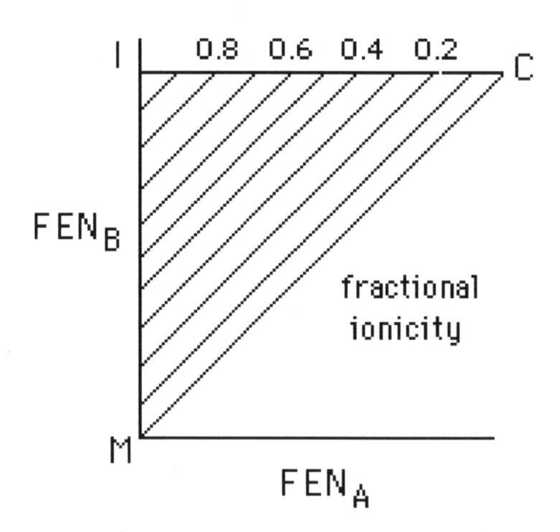

equilateral triangles (19-20).

Bond	*Fractional Ionicity*	*Fractional Covalency*	*Fractional Metallicity*	*Sum*
Cs-F	1.00	0.00	0.00	1.00
Cs-Cs	0.00	0.00	1.00	1.00
F-F	0.00	1.00	0.00	1.00
Ba-F	0.90	0.10	0.00	1.00
Mg-Al	0.10	0.19	0.71	1.00
Cl-O	0.11	0.74	0.15	1.00
Ge-I	0.31	0.31	0.38	1.00

As an illustration of the application of equations 1-4, the above table summarizes the calculated fractional bond coordinates for several representative bonds in both versions of the bond triangle. Of course, if one prefers, one can express these coordinates as percentages rather than as fractions and thus talk of the Ge-I bond, for example, as being 31% ionic, 31% covalent and 38% metallic.

It is important to emphasize that the above values of fractional bond character are internally defined by the bond triangle itself, and are not the result applying various external criteria, however circular, as was the case with the discrete classes assumed by Sproul and Allen. To say that the Si-H bond is 10% ionic, 40% covalent and 50% metallic, whereas the Si-F bond is 60% ionic, 40% covalent and 0% metallic is mathematically equivalent to displaying the relative positions of the two bonds on the bond triangle. The latter merely presents the relationship visually, whereas the former does so by explicitly listing their relative fractional distances from each of the three vertices. If one rejects the fractional bond coordinates as unfounded, then one must also reject the use of the bond triangle itself.

4.4 Predominant Bond Type

One of the forces which inhibits the full pedagogical implementation of the continuum interpretation of bond-type triangles is the ever-present pressure on teachers and textbook authors to reduce chemical concepts to black and white, either/or, categories in order to facilitate student memorization, problem solving and test taking. This problem can be addressed in manner uniquely different from that assumed by the separation postulate through the use of triangular coordinates and the resulting concept of predominant bond-type. With the exception of the single point in the center of the bond triangle, having the fractional bond coordinates of 33% ionic, 33% covalent, and 33% metallic, and the midpoints or 50% points of the three edges, all other points within the triangle must of necessity lie closer to one of the vertices than to the other two and this fact may be used to define its predominant bond type. Connecting these four unique points with one another divides the triangle into the three regions shown on the following page.

All bonds lying in the upper region will automatically have a fractional ionicity value greater than either their fractional covalency or their fractional metallicity values and may therefore be described as being "predominantly ionic" in character. Likewise all bonds lying in the lower

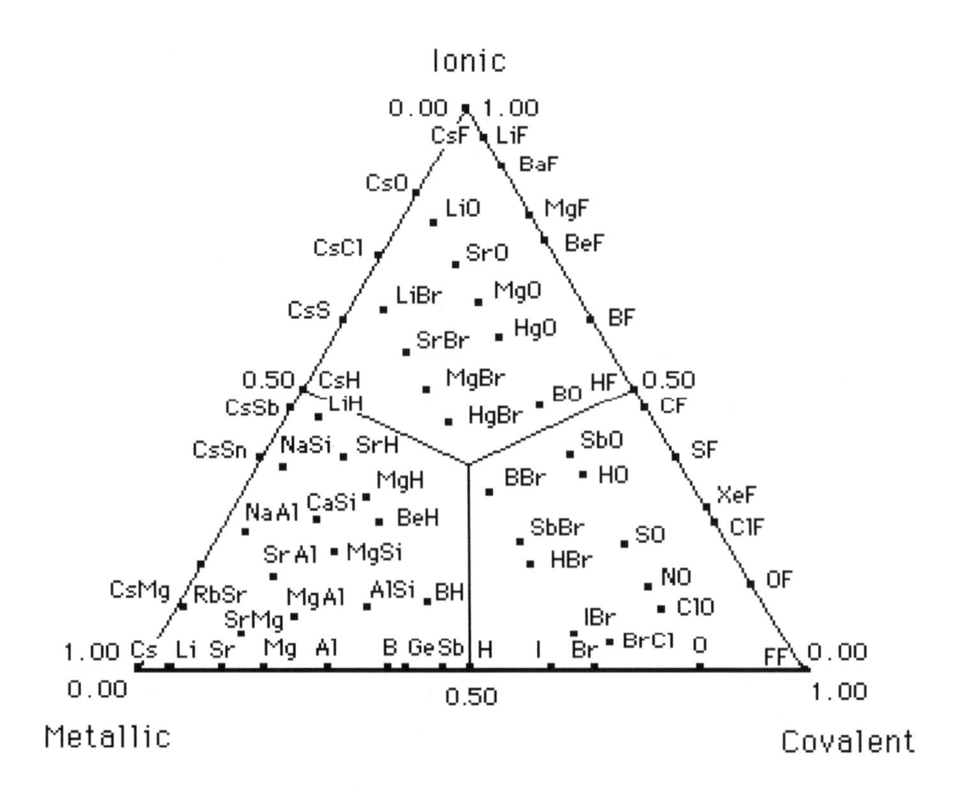

right-hand region will automatically have a fractional covalency value greater than either their fractional ionicity or their fractional metallicity values and may therefore be described as being "predominantly covalent" in character. Finally, all bonds lying in the lower left-hand region will automatically have a fractional metallicity value greater than either their fractional ionicity or fractional covalency values and may therefore be described as being "predominantly metallic" in character.

Note once again that these terms simply refer to relative positions within the bond triangle and fact that the bond in question is closer to a particular vertex than to the other two. Unlike the regions in the Sproul-Allen diagrams, they under no circumstances imply that one would expect to see a sudden and discontinuous change in the properties of the resulting compounds and simple substances on passing from one region to another. Further note that, unlike the divisions within the Sproul-Allen diagrams, none of these boundaries are parallel to the sides of the triangle. Consequently they do not automatically imply that all of the bonds of a given element, whether lying along either its electropositive or its electronegative branches, must of necessity fall into the same category. Thus they classify the Si-H bond as predominantly metallic and the Si-F as predominantly ionic rather than requiring that both be equally borderline ionic. Nor, as already emphasized, do these labels imply that the silanes automatically have the properties of bulk metals and the silicon fluorides the properties of ionic salts, but rather only that the Si-H bond lies closer to the metallic vertex and is rather diffuse, whereas the Si-F bond lies closer to the ionic vertex and is relatively polar.

Of course, the question naturally arises as to how well these definitions of fractional bond character and predominant bond type correlate with truly independent external empirical and theoretical criteria thought to be characteristic of bond type. As we will see in the next section, when it comes to empirical criteria, the relationship between bond type and the properties of the resulting compounds and simple substances is far more complicated than is usually assumed by textbook authors. Likewise, there are many competing theoretical definitions of bond type in the chemical literature, the vast majority of which are based on the assumption of only two (ionic and covalent), rather than three limiting cases, thus complicating their direct comparison with the above definitions (21). Nevertheless some suggestive results based on empirical sorting maps will be presented in Lecture IX. For the present, however, we will simply take advantage of the fact that these definitions allow us to speak in a more quantitatively precise manner of just what we mean by the relative ionicity, covalency and metallicity of various bonds, whatever may ultimately prove to be their relationship (or lack thereof) to such external criteria.

4.6 Bonds versus Compounds and Simple Substances

Despite the fact that they are called "bond-type" triangles, our historical survey of the literature on this subject in Lecture II has shown that there is considerable ambiguity over whether the data points in these triangles are intended to represent bonds or whether they are intended to represent actual compounds and simple substances and, if the latter, whether these are merely specific examples of substances containing the bonds in question or one is claiming that it is the bulk substances themselves which are being classified as ionic, covalent, or metallic. This latter interpretation is certainly implied by the separation postulate of Yeh, Sproul and Allen and is a natural consequence of the widespread belief that simple substances and binary compounds may be meaningfully classified solely on the basis of their component bond types, as well as the further belief that the resulting classification correlates in a significant fashion with such bulk properties of the substances as degree of polymerization, melting point, solubility in polar versus nonpolar solvents, hardness, degree of ionic dissociation in solution, and degree of metallic conductivity.

Such a simple one to one relationship between bond type and bulk properties assumes at the very least that the substances in question contain only one kind of bond. If it is really the substances themselves which are being plotted, then this requirement would automatically restrict application of the triangle to a survey of simple substances and binary compounds only and, indeed, to only a small fraction of the binary compounds presently known. This is because, as we will see in greater detail in the following lectures, the majority of known binary compounds contain not only A-B bonds but also A-A and/or B-B bonds as well – a fact which, for compounds of this type, would automatically preclude the possibility of a simple one to one relationship between bulk properties and A-B bond type alone.

In order to make this problem more explicit, we may divide binary compounds into the two large classes of "homodesmic" (from the Greek *homos*, meaning "alike" and *desmos*, meaning "to bind or bond") and "heterodesmic" (from the Greek *heteros*, meaning "different" and *desmos*, meaning "to bind or bond") (22). As these terms imply, homodesmic or simple binaries contain only one kind of bond – the A-B bond – and only they qualify as likely candidates for a one to one correlation between bond type and bulk properties, whereas heterodesmic binaries obviously do not. Heterodesmic binaries may, in turn, be further subdivided into three classes: polycationic binaries, containing both A-B and A-A bonds; polyanionic binaries, containing both A-B and B-B bonds; and mixed binaries containing A-B, A-A and B-B bonds (23):

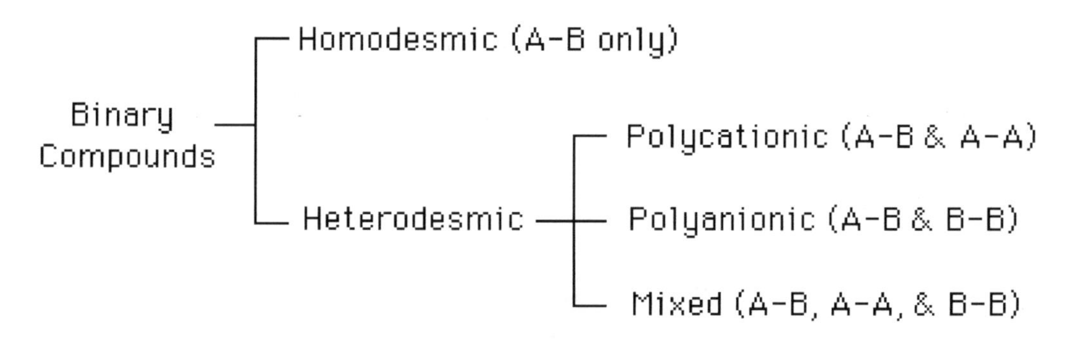

As before, the terms cationic and anionic are used in a purely formal sense as synonyms for the electropositive and electronegative components of the compounds and bonds. Likewise, the terms A-A bond and B-B bond refer to true interatomic covalent and/or metallic bonding and not to weak intermolecular forces between like atoms.

In the case of polycationic binaries, the A-A bonding network may correspond to discrete chains, such as the Hg-Hg bonded dimers in Hg_2Cl_2; to discrete rings or polycycles, such as the B-B bonded tetramers in B_4Cl_4; to infinitely extended chains, such as the Ni-Ni bonded chains in NiAs; to infinitely extended layers, such as the C-C bonded layers in CF; or to infinitely extended frameworks, such as the Ti-Ti bonded framework in TiO.

In the case of polyanionic binaries, the B-B bonding network may likewise correspond to discrete chains, such as the I-I bonded trimers in CsI_3; to discrete rings or polycycles, such as the Pb-Pb bonded tetramers in $K_4(Pb_4)$, to infinitely extended chains, such as the P-P bonded chains in ZnP_2, to infinitely extended layers, such as the Si-Si bonded layers in $CaSi_2$, or to infinitely extended frameworks, such as the Tl-Tl bonded framework in NaTl. The compound HgN_3 is a good example of a mixed heterodesmic binary as it contains, in addition to the Hg-N bonds, both finite Hg-Hg bonded chains and finite N-N bonded (azide) chains.

As will be detailed in the following four lectures, polycationic and polyanionic binaries are relatively common – a fact which is seldom acknowledged by introductory textbooks, which tend instead to focus almost exclusively on homodesmic or simple binaries and polycationic binaries containing cationic lone pairs. This neglect is further accentuated by the common practice of using net stoichiometric formulas to describe these species – a practice which effectively hides their underlying structural complexity (24).

Though a homodesmic bonding system is a necessary requirement for a one to one relationship between bond type and bulk properties, it is by no means sufficient (25). Thus the common textbook claim that typical ionic substances are nonmolecular and possess high melting points, whereas typical covalent substances are molecular and possess low melting points, is contradicted by the such examples as diamond and tungsten, which contain pure covalent and metallic bonds, respectively, but which are both non-molecular solids having extremely high melting points (3547 °C and 3407 °C respectively), and by the example of SF_6, which contains fairly ionic bonds, but which is a discrete molecule with a low melting point (-50.7 °C). Likewise, the textbook claim that only compounds with ionic bonds undergo ionic dissociation on dissolving in polar solvents is contradicted by the fact that the covalently bonded I_2 molecule dissolves in relatively nonpolar pyridine (C_6H_5N) to form a solution possessing significant ionic conductivity:

$$C_6H_5N(l) + 2I_2(s) \rightarrow C_6H_5NI^+(sol) + I_3^-(sol) \quad [5]$$

In fact, as was first emphasized by Kossel in 1920 (26) and again by Pauling in 1932 (27), by Stillwell in 1938 (28), by Lingafelter in 1993 (29), and more recently by the current author (30, 31), sudden and discontinuous changes in such bulk properties as melting point almost always reflect a sudden and discontinuous change in structure type rather than a sudden and discontinuous change in bond type. Consequently compounds and simple substances are best described using a classification based explicitly on structure (i.e., discrete molecular, infinite framework, infinite layer, and infinite chain) rather than one based on bond type alone (i.e., ionic, covalent and metallic). This structural classification has been used by various authors in the past (32) and may be conveniently summarized by means of a tetrahedral diagram, first introduced by Grimm in 1934 (33), and shown in the following updated figure (30, 31):

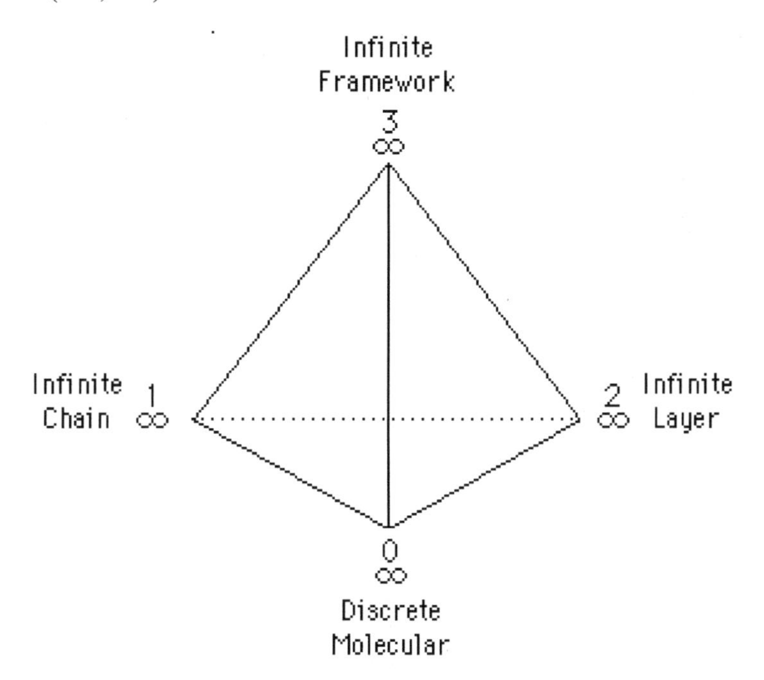

Of course, claiming that structure plays the dominant role in determining bulk properties is not the same as claiming that bond type plays no role whatsoever. Rather the true situation is best represented by a function of the form:

$$\textit{bulk properties} = f(\textit{bond type, structure}, P, T) \qquad [6]$$

which recognizes the influence of both, as well as the long-established roles of temperature (*T*) and pressure (*P*). Such a function implies that any hope of truly dissecting out the role of bond type alone would require a comparison of a series of isostructural substances at constant *T* and *P*. An example is the correlation between Δ*EN* and the bulk modulus for alkali halides having a common 6/6 framework structure (34). The question of how to integrate the effects of both bond type and structure into a single diagram will be discussed in Lecture 10.

To summarize, the simple bond triangle is, as its name implies, a classification of bonds and not of substances, and the classification of the latter is best served instead by a classification based on structure rather than bond type. In order to minimize the confusion of these two subjects it is best to never to label the data points within the bond triangle using the formulas of example compounds and simple substances, but rather to always use labels indicating only the component atoms forming the bond – in other words to label the points simply as AB, as in the example triangles given in Lecture III and in Section 4.4, or better still, to label them simply as A-B.

4.6 Atoms versus Molecules

It was pointed out in Section 3.1 that, because of the high degree of linear correlation displayed by the 25 or more *EN* scales in the chemical literature, it makes little practical difference which scale is selected to plot the triangle, provided that it is complete. Nor does it make any difference whether one uses the absolute *EN* values for the scale in question, their Pauling adjusted values, or their corresponding reduced *FEN* values. In the first instance one obtains the same general overall distribution of points, and in the second instance the identical relative distribution. However, when it comes to interpreting the physical significance of the coordinates, a consideration of the theoretical basis of the *EN* scale of choice becomes important.

More than 40 years ago Ferreira suggested that one should distinguish between what he called primary and secondary *EN* scales (35):

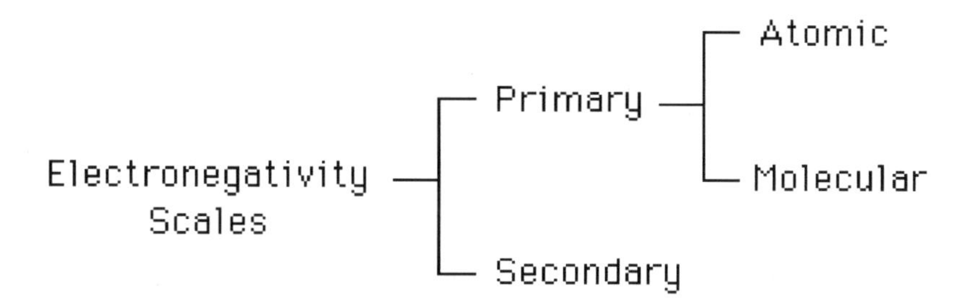

A primary scale is one which directly measures either the force or the energy of attraction of an atom for its valence electrons. It is often calculated using such experimental properties as ionization potentials, electron affinities, radii, effective nuclear charges, valence-electron counts, orbital energies, etc. thought to directly measure this force or energy.

A secondary scale, on the other hand, is derived indirectly on the basis of an empirical correlation between either *EN* values or *ΔEN* values and some molecular property thought to parallel these parameters, such as heats of formation, vibrational force constants, dipole moments, spectroscopic charge-transfer maxima, and K_{sp} values, in the case of *ΔEN*, or solid-state work functions and polarizabilites, in the case of *EN* alone. Generally one establishes the correlation using a few known *EN* values, assumes that it is universal, and then recalculates both the initial *EN* values, as well as additional *EN* values not used in the original correlation, by assuming that the correlation is perfect.

Since correlations of this type are often lacking a rigorous theoretical basis and may be the result of a fortuitous cancellation of various complicating factors, such secondary scales are to be avoided in constructing bond triangles. This advice is offered despite the fact that the Pauling thermochemical *EN* scale, which is the granddaddy of all such secondary scales, continues to exercise enormous influence among textbook authors. Not only does it continue to function as the numerical "reference gauge" against which all other primary and secondary scales measure their success, most of these alternative scales have also been forced, in order to win some degree of acceptance, to adjust their absolute *EN* values so as to conform as closely as possible to the purely arbitrary numerical range (0.5 - 4.0) originally adopted by Pauling for his secondary scale.

Primary *EN* scales may be further subdivided into the classes of atomic and molecular. Primary atomic *EN* scales, as the name implies, are a property of isolated gas-phase atoms only. Examples include the Allred-Rockow electrostatic force definition, the Sanderson stability ratio scale, the Allen spectroscopic or configuration energy scale, and the Bohr average ionization-xenergy scale discussed in Appendix I.

Primary molecular *EN* scales, on the other hand, are atomic *EN* values which have been modified to accommodate the various electronic perturbations which occur when an atom is incorporated into a molecule. In other words, rather than measuring the *EN* of the isolated atom they instead attempt to measure the *EN* of the "atom" within the molecule. Thus attempts to accommodate perturbations produced by changes in the kinds of adjacent ligands (i.e., the *EN* value of C relative to I in $I\text{-}CH_3$ versus $I\text{-}CF_3$) have given rise to various "group" electronegativity schemes, such as those of summarized by Wells (36). Likewise, attempts to accommodate perturbations produced by changes in the number and geometrical arrangement of the adjacent ligands (i.e. the *EN* of P in PCl_3 versus PCl_5 or that of sp^3 - versus sp^2- versus *sp*-bonded C) have given rise to various hybridization-dependent *EN* schemes, such as the Mulliken definition, or to various charge-dependent schemes, such as the Iczkowski-Margrave definition, or both, as in the Hinze-Jaffe orbital definition. More recently Bratsch has shown how the latter definition can be used to calculate charge- and hybridization-dependent group *EN* values as well, thus producing a composite scheme which is sensitive to inductive effects, as well as to charge-transfer and hybridization effects (37). All of these schemes rely heavily on the postulate of electronegativity neutralization and have, in effect, made the *EN* of the atom within the molecule an explicit function of the molecule's overall composition and structure.

The problem is that, though these elaborations are able to provide more information than simple atomic *EN* values alone, they also require additional input data concerning a molecule's bonding topology and structure – a feature which limits their application. Indeed, though these elaborated molecular *EN* schemes have been around for almost 50 years now, they are almost never applied in detail in textbooks, suggesting that the perceived advantages are offset by their increased complexity. Rather most textbook continue to treat *EN* as an atomic property which uniquely characterizes each atom type and employ it to establish broad correlations between certain properties of molecules and periodic trends in the *EN* values of their component atoms.

This conflict is a classic example of the old conundrum that breath of treatment usually requires the sacrifice of depth of treatment and vice versa:

$$(breadth)(depth) = constant \qquad [7]$$

If the bond-type triangle is to have any value it must have great breath of application. We are looking for sweeping, albeit crude, correlations between bond-type in the broadest possible range of bonding situations and some property of the isolated atoms forming the bond. This mandate requires that we employ a primary atomic *EN* scale, rather than a primary molecular *EN* scale – whence our decision to use the Bohr average ionization-energy scale in Appendix 1 throughout this monograph.

4.7 A Compromise Approach

In light of the above discussion, function 6 is in fact not very useful, since, in reality, bond type and structure type are not truly independent of one another but are rather functions of many of the same underlying factors. A more fruitful and more reductionistic approach is treat the properties of a substance (including both its bond type and structure type) as a function of the electronegativities of its component atoms (FEN_A, FEN_B), the ratio in which the atoms are combined (b/a), the overall size of the valence manifolds or block assignments of the component atoms (m), and both temperature (T) and pressure (P):

$$properties = f(FEN_A, FEN_B, b/a, m, P, T) \quad [8]$$

From this point of view, a bond-type triangle, which seeks to correlate bond type with atomic electronegativities alone, is subject to the constraint that the other four parameters should be kept as constant as possible:

$$bond\ type = f(FEN_A, FEN_B)_{b/a,\, m,\, P,\, T} \quad [9]$$

This is easy enough in the case of P and T since it is always implicitly assumed that we are dealing with simple substances and compounds at room temperature and pressure (RTP). Likewise, in the case of m we need only avoid the mixing H-He, main-block, and transition-block species on the same diagram. Rather it is the constant b/a constraint that is the most difficult to maintain.

As we will see in the following four lectures, if we survey our bonds within the main block on a group by group basis, then, for the case of possible homodesmic compounds, this latter condition is in fact met. Unhappily, as we will also see, many of these potential homodesmic compounds are at present unknown and one must instead resort to various polycationic and polyanionic heterodesmic examples for which this constraint is no longer rigorously maintained. In Lecture IX, we will further explore how function 9 may be applied within the context of the bond-type triangle to map variations in various other properties, such as structure type, band-gap energies, color, and bulk electrical conductivities.

4.8 References and Notes

1. P. Y. Yeh, "A Chart of Chemical Compounds Based on Electronegativity," *J. Chem. Educ.*, **1956**, *33*, 134.

2. G. D. Sproul, "Electronegativity and Bond Type: 1. Tripartite Separation," *J. Chem. Educ.*, **1993**, *70*, 531-534. Inspection of the 311 binaries used by Sproul (explicitly listed in reference 5) shows that 73 or 23% of them are redundant, i.e., that they are alternative examples for the same bond (e.g. CH_4 versus C_2H_6 for the C-H bond).

3. L. C. Allen, J. F. Capitaini, G. A. Kolks, G. D. Sproul, "Van Arkel - Ketelaar Triangles," *J. Mol. Struct.*, **1993**, *300*, 647-655.

4. A. F. Wells, *Structural Inorganic Chemistry*, 5th ed., Clarendon: Oxford, 1984, Chapter 7.

5. G. D. Sproul, "Electronegativity and Bond Type: 2. Evaluation of Electronegativity Scales," *J. Phys. Chem.*, **1994**, *98*, 6699-6703.

6. G. D. Sproul, "Electronegativity and Bond Type: 3. Origins of Bond Type," *J. Phys. Chem.*, **1994**, *98*, 13221-13224. In this paper and again in reference 7 Sproul presents not only his quantification of the Fernelius triangle but also a quantified version of Yeh's orientation of the Grimm triangle without acknowledgement of Yeh's priority. Sproul's further failure to appreciate the fact that these two versions of the bond-triangle are related by a simple coordinate transformation that does not alter the relative positions of the points within the triangle also leads to an erroneous claim that adoption of the Yeh version in place of the Fernelius version decreases the percentage of errors for his various boundaries from 16% to 4% or nearly fourfold.

7. G. D. Sproul, "Electronegativity and Bond Type. Predicting Bond Types," *J. Chem. Educ.*, **2001**, *78*, 387-390.

8. L. C. Allen, "Electronegativity is the Average One-Electron Energy of the Valence-Shell Electrons in the Ground-State Free Atom." *J. Am. Chem. Soc.*, **1989**, *111*, 9003-9014, footnote 2. See also note 21 of Lecture II.

9. E. G. Rochow, *The Metalloids*, D. C. Heath: Boston, MA, 1966, p. 7.

10. A fivefold division of the bond triangle into discrete ionic, covalent, metallic, polar covalent and semimetallic regions using, in some cases, nonlinear boundaries has

more recently been proposed in T. L. Meek, L. D. Gardner, "Electronegativity and the Bond Triangle," *J. Chem. Educ.*, **2005**, *82*, 325-333. However, these authors also use a modified set of coordinates. Consequently their proposal will not be discussed until Lecture 11.

11. G. M. Bodner, L. H. Richard, J. N. Spencer, *Chemistry: Structure and Dynamics*, Wiley: New York, NY, 1996, pp. 186-189.

12. G. Wulfsberg, *Inorganic Chemistry*, University Science Books: Sausalito, CA, 2000, pp. 775-778.

13. G. Raynor-Canham, T. Overton, *Descriptive Inorganic Chemistry*, Freeman, New York, NY, 2006, pp. 102-103.

14. Thus reference 11 incorrectly reproduces Allen's metalloid region by using the electropositive branch for Al rather than for Si to determine the ionic-covalent boundary. The resulting semimetal or SM region, as these authors prefer to call it, contains many elements not included in their earlier definition of semimetals (p. 17) – a definition which in fact coincides with the Rochow metalloid definition used by Allen.

15. For a general historical account of ternary diagrams, see R. J. Howorth, "Sources for a History of the Ternary Diagram," *Brit. J. Hist. Sci.*, **1996**, *29*, 337-356.

16. Older textbooks on the phase rule generally provide a more explicit introduction to the mathematics of these diagrams. For a good example, see J. S. Marsh, *Principles of Phase Diagrams*, McGraw-Hill: New York, NY, 1935, Chapter IV.

17. R. M. Evans, *An Introduction to Color*, Wiley: New York, NY, 1948, pp. 114-115.

18. W. Vogel, *Chemistry of Glasses*, American Ceramics Society: Columbus, OH, 1985, pp. 110, 137, 184-191. Also O. V. Mazurin, M. V. Streltsina, and T. P. Shvaiko-Shvaikovskaya, *Handbook of Glass Data, Part C: Ternary Silicate Glasses*, Elsevier: Amsterdam, 1987. Ternary compositions which form glasses are marked on these diagrams which superficially look like equilibrium phase diagrams, though glasses, by definition, are nonequilibrium or metastable phases.

19. For a general discussion of alternative graphical representations of the composition of ternary systems, see F. G. Smith, *Physical Geochemistry*, Addison-Wesley: Reading, MA, 1963, pp.128-132.

20. For specific examples of right-triangular ternary phase diagrams, see G. G.

Brown, Ed., *Unit Operations*, Wiley: New York, NY, 1950, pp. 287-292, 303-305. For an example of a right-triangular Maxwell color diagram, see reference 17, p. 115.

21. For an overview of various alternative definitions of bond ionicity, see J. Barbe, "Convenient Relations for the Estimation of Bond Ionicity in A-B Type Compounds," *J. Chem. Educ.*, **1983**, *60*, 640-642.

22. This terminology is based on that used in R. C. Evans, *An Introduction to Crystal Chemistry*, Cambridge University Press: Cambridge, 1948, Chapter 7, which was, in turn, apparently based on the suggestions of J. D. Bernal.

23. U. Müller, *Inorganic Structural Chemistry*, Wiley: New York, NY, 1993. Chapter 12.

24. Net stoichiometric formulas may in fact be further elaborated so as make explicit the various polycationic and polyanionic substructures present in heterodesmic binaries. For details see W. B. Jensen, "Crystal Coordination Formulas: A Flexible Notation for the Interpretation of Solid-State Structures," in D. G. Pettifor, Ed., *The Structures of Binary Compounds*, North-Holland: Amsterdam, 1989, pp. 105-146.

25. In this regard it is of interest to note that only a small percentage of the species used by Sproul to establish his discrete ionic, covalent and metallic boundaries are homodesmic.

26. W. Kossel, "Bemerkungen über Atomkrafte," *Zeit. Phys.*, **1920**, *1*, 395-415.

27. L. Pauling, "The Nature of the Chemical Bond. III. The Transition from One Extreme Bond to Another," *J. Am. Chem. Soc.*, **1932**, *54*, 988-1003.

28. C. W. Stillwell, *Crystal Chemistry*, McGraw-Hill: New York, NY, 1938, pp. 165-177.

29. E. C. Lingafelter, "Why Low Melting Doesn't Indicate Covalency in MX_4 Compounds," *J. Chem. Educ.*, **1993**, *70*, 98-99.

30. W. B. Jensen, "Logic, History and the Chemistry Textbook. II. Can We Unmuddle the Chemistry Textbook?," *J. Chem. Educ.*, **1998**, *75*, 817-828.

31. W. B. Jensen, "Bond Type versus Structure Type," *Educ. Chem.*, **1994**, *31*, 94.

32. Example textbooks using this structural approach include G. Champetier, *Les molécules géantes et leurs applications*, Albin Michel: Paris, 1948, Part II, Chapter 1; P. D. Ritchie, *A Chemistry of Plastics and High Polymers*, Cleaver-Hume: London, 1949, Chapter 14; H. J. Emeléus, J. S. Anderson, *Modern Aspects of Inorganic Chemistry*, Van

Nostrand, New York, NY, 1952, p. 71; A. F. Wells, *Models in Structural Inorganic Chemistry*, Oxford University Press: Oxford, 1970, p. 7; W. L. Jolly, *Modern Inorganic Chemistry*, McGraw-Hill: New York, NY, 1984, Chapter 11.

33. H. G. Grimm, "Allgemeines über die verschiedenen Bindungsarten," *Z. Elekrochem.*, **1928**, *34*, 430-437.

34. J. Gilman, *Electronic Basis of the Strength of Materials*, Cambridge University Press: Cambridge, 2003, p. 137. This book contains many isostructural plots for AB species.

35. R. Ferreira, "Electronegativity and Chemical Bonding," *Adv. Chem. Phys.*, **1967**, *13*, 55-84.

36. P. R. Wells, "Group Electronegativities," *Prog. Phys. Org. Chem.*, **1969**, *6*, 111-145.

37. S. G. Bratsch, "Revised Mulliken Electronegativities." *J. Chem. Educ.*, **1988**, *65*, 34-41, 223-227.

V

Bonding in the Binary Compounds of Group *1/7*

5.1 Group Labels and the Periodic Table

Our intent in this and the following three lectures is to survey the possible binary compounds formed between the various the main-block elements in order to illustrate how the Fernelius version of the bond-type triangle can be used to characterize their resulting A-B bonds. However, before proceeding with our survey, it is necessary to first provide some brief background about both the specific form of periodic table we have chosen to employ and the accompanying labeling system for the various groups. The table in question, which is shown on the following page, first divides the known elements into four fundamental classes or blocks based on the overall capacity of their atoms' valence manifolds: the hydrogen-helium block (maximum valence capacity of 2), the main block (maximum valence capacity of 8), the transition block (maximum valence capacity of 18), and the inner-transition block (maximum valence capacity of 32) (1). The table does not make use of the usual s- and p-subdivisions of the main block since, in reality, both so-called s-block elements and so-called p-block element make use of the full range of available filled and empty outer s- and p-orbitals in their bonding interactions.

Thus, for the hydrogen-helium block elements, the sum of the various group valence electron (*e*) and valence vacancy (*v*) counts always sum to two:

$e + v = 2$ *(hydrogen-helium block)* [1]

H-He Block
(e + v = 2)

	1/1	*2/0*
1	H	He

Main Block
(e + v = 8)

	1/7	*2/6*	*2/6'*	*3/5*	*4/4*	*5/3*	*6/2*	*7/1*	*8/0*
2	Li	Be		B	C	N	O	F	Ne
3	Na	Mg		Al	Si	P	S	Cl	Ar
4	K	Ca	Zn	Ga	Ge	As	Se	Br	Kr
5	Rb	Sr	Cd	In	Sn	Sb	Te	I	Xe
6	Cs	Ba	Hg	Tl	Pb	Bi	Po	At	Rn
7	Fr	Ra	112	113	114	115	116	117	118

Transition Block
(e + v) = 18)

	3/15	*4/14*	*5/13*	*6/12*	*7/11*	*8/10*	*9/9*	*10/8*	*11/7*
4	Sc	Ti	V	Cr	Mn	Fe	Co	Ni	Cu
5	Y	Zr	No	Mo	Tc	Ru	Rh	Pd	Ag
6	Lu	Hf	Ta	W	Re	Os	Ir	Pt	Au
7	Lr	Rf	Db	Sg	Hs	Mt	109	110	111

Inner-Transition Block
(e + v = 32)

	3/29	*4/28*	*5/27*	*6/26*	*7/25*	*8/24*	*9/23*	*10/22*	*11/21*	*12/20*	*13/19*	*14/18*	*15/17*	*16/16*
6	La	Ce	Pr	Nd	Pm	Sm	Eu	Gd	Tb	Dy	Ho	Er	Tm	Yb
7	Ac	Th	Pa	U	Np	Pu	Am	Cm	Bk	Cf	Es	Fm	Md	No

For the main-block elements, they always sum to eight:

$$e + v = 8 \quad (main\ block) \qquad [2]$$

For the transition block elements, they always sum to eighteen:

$$e + v = 18 \quad (transition\ block) \qquad [3]$$

and for the inner-transition block elements, they always sum to thirty two:

$$e + v = 32 \quad (inner\text{-}transition\ block) \qquad [4]$$

Each of the groups within these four fundamental blocks is assigned a dual numerical label, the numerator of which denotes the number of valence electrons (e) present in the valence manifolds of the various atoms belonging to the group in question, and the denominator of which denotes the number of valence vacancies (v) present in their valence manifolds. Thus hydrogen is in Group *1/1* (one valence electron, one vacancy), boron is in Group *3/5* (three valence electrons, five vacancies), Cr is in Group *6/12* (six valence electrons, twelve vacancies) and Ce is in Group *4/28* (four valence electrons, 28 vacancies), etc.

Note that not only do these e/v labels characterize an atom's electron-donor properties, like most other group labels, they also characterize its electron-acceptor properties. Both of these parameters play equally important roles in determining the chemical behavior of an atom and the traditional focus on the former at the expense of the latter has been the source of many misunderstandings about the periodic table. For example, this block structure and notational system unambiguously resolves the perennial debate over whether H belongs to the alkali metals (Group *1/7*) or to the halogens (Group *7/1*), as well as the more recent fuss over whether He belongs to the alkaline-earth metals (Group *2/6*) or to the noble gases (Group

8/0), as it clearly shows that none of these options is valid. In actual fact, as already indicated, H and He belong to a block of their own and their resulting group assignments (Groups *1/1* and *2/0* respectively) clearly indicate that they are not isovalent (both identical valence electron and valence vacancies counts) with any of the above main-block groups, though they do show both weaker secondary inter-block relationships to Groups *1/7* and *2/6* by virtue of identical valence electron counts, but nonidentical valence vacancies accounts, and weaker tertiary inter-block relationships to Groups *7/1* and *8/0* by virtue of identical valence vacancies counts, but nonidentical valence electron counts.

5.2 Binary Compounds and the Periodic Table

In the following survey we will first give a tabular summary the known binary compounds formed between the alkali metals or group *1/7* elements and each of the remaining groups of the main-block, beginning with the halogens or group *7/1* elements. For each combination of groups the table of known binary compounds will then be followed by a bond-type triangle indicating the location of the resulting A-B bonds. Each box in the table of binary compounds will list the net compositional formulas, A_aB_b, of the known binary compounds for the particular AB element combination in question in order of the increasing value of the ratio (*b/a*) of their stoichiometric coefficients – an order which also corresponds to the sequence: heterodesmic polycationic binaries, homodesmic simple binaries, and heterodesmic polyanionic binaries. In each case the formula for the homodesmic compound, which marks the midpoint in this sequence, will also be underlined for quick identification.

In each case, the value of the *b/a* ratio for the formula of the single unique simple homodesmic binary may be easily predicted from the positions of its component A and B atoms in the periodic table and the *e/v* group labeling scheme introduced in the previous section. For these compounds the total number of valence electrons (ae) provided by the electropositive A atoms must be equal to the total number of valence vacancies (bv) available on the electronegative B atoms:

$$ae_A = bv_B \qquad [5]$$

Hence the ratio of the resulting stoichiometric coefficients is given by the relationship:

$$b/a = e_A/v_B \qquad [6]$$

or, in other words:

$$A_aB_b = A_{vB}B_{eA} \qquad [7]$$

Though this simple calculation is often used in textbooks to predict the stoichiometries of ionic compounds, it use does not preclude the possibility that resulting A-B bonds may actually be polar covalent or polar metallic in nature. When $b/a < e_A/v_B$, there are too many valence electrons available on the electropositive atoms and these extra electrons may either be used for the formation of covalent or metallic A-A bonds or cationic lone pairs. Likewise, when $b/a > e_A/v_B$, there are too many valence vacancies available on the electronegative atoms, and these extra vacancies must be filled instead via the formation of covalent or metallic B-B bonds.

As we will see below, polycationic and polyanionic heterodesmic binaries are relatively common – a fact that is seldom acknowledged by introductory textbooks, which tend instead to focus almost exclusively on simple homodesmic binaries and polycationic binaries containing cationic lone pairs. However, as one approaches the metallic corner of the bond-type triangle, in particular, polyanionic and polycationic binaries containing B-B and A-A bonds, as well as A-B bonds, become the rule rather than the exception. Since one has more than one kind of bond present, the simple assumption that not only the A-B bonds, but the resulting compounds themselves, can be unambiguously classified as ionic, covalent or metallic breaks down. Thus, in the case of polyanionics, one may have both a transition of the A-B bond from ionic to metallic and a simultaneous transition of the B-B bonds from covalent to metallic. Likewise, in the case of polycationics, one often sees a transition of the A-A bonds from metallic to covalent superimposed on a simultaneous transition of the A-B bonds from ionic to covalent. Two consequences of this are that so-called intermetallic compounds often exhibit bulk semiconductor rather than bulk metallic properties for a considerable distance into the predominantly metallic region of the triangle and that the resulting ELF localization studies of their electron density distributions often reveal a complex mixture of bonding situations (2). Thus, in the following four lectures, it is important to emphasize that we are tracking the progressive changes in the A-B bond only and not those of any A-A or B-B bonds which might also be simultaneously present, and that our use of the terms predominantly ionic, predominantly covalent, and predominantly metallic refers only to the characterization of these A-B bonds and not to the compound as a whole.

5.3 Group *1/7* - Group 7/1 Binaries

As already indicated, we begin our survey in this lecture with the binary compounds formed between the alkali metals or Group *1/7* elements and the halogens or Group *7/1* elements, since, at present, there is no evidence for compound formation between the alkali metals and the noble gases or elements of Group *8/0* (3). In addition, we will also exclude the compounds of the radioactive elements Fr, Ra, Po, At, and Rn from our survey of binary compounds, as their chemistry is often very imperfectly known, though their potential bonds will be included in the corresponding bond-type triangles. These same restrictions will be applied in the following three lectures as well.

The net stoichiometries of most of the known Group *1/7* - Group *7/1* binary compounds are summarized in the following table (4):

B \ A	Li	Na	K	Rb	Cs
F	LiF	NaF	KF	RbF	CsF
Cl	LiCl	NaCl	KCl	RbCl	CsCl
Br	LiBr	NaBr	KBr	RbBr	CsBr $CsBr_3$
I	LiI	NaI	KI KI_3	RbI RbI_3	CsI CsI_3 Cs_2I_4 Cs_2I_8

From relation 6 the stoichiometry for the simple homodesmic binaries is predicted to be AB. Thus for homodesmic binaries: $b/a = e_A/v_B = 1/1$, for polyanionic heterodesmic binaries: $b/a > 1/1$, and for poly-cationic heterodesmic binaries: $b/a < 1/1$. As may be seen, all of the resulting A-B bonds may be illustrated using highly stable, simple homodesmic binaries and, in the case of Rb and Cs, also in the form of several stable polyanionic salts of the composition $A(B_b)$ containing various polyiodide chain anions such as I_3^- (5). Though alkali metal compounds containing polyanions of the other halogens may also be prepared under extreme conditions, they are highly unstable.

In keeping with the standard textbook characterization of these A-B bonds as ionic, we see that all of them are located within the parallelogram shown in the upper left-hand corner of the following bond-type triangle, whose boundaries are defined by Ce-F, Li-F, Cs-I, Li-I, and that indeed they all fall within the predominantly ionic region with their numerical ionic characters ranging from a maximum of 100% by definition for the Cs-F bond to a minimum of 57% for the Li-I bond:

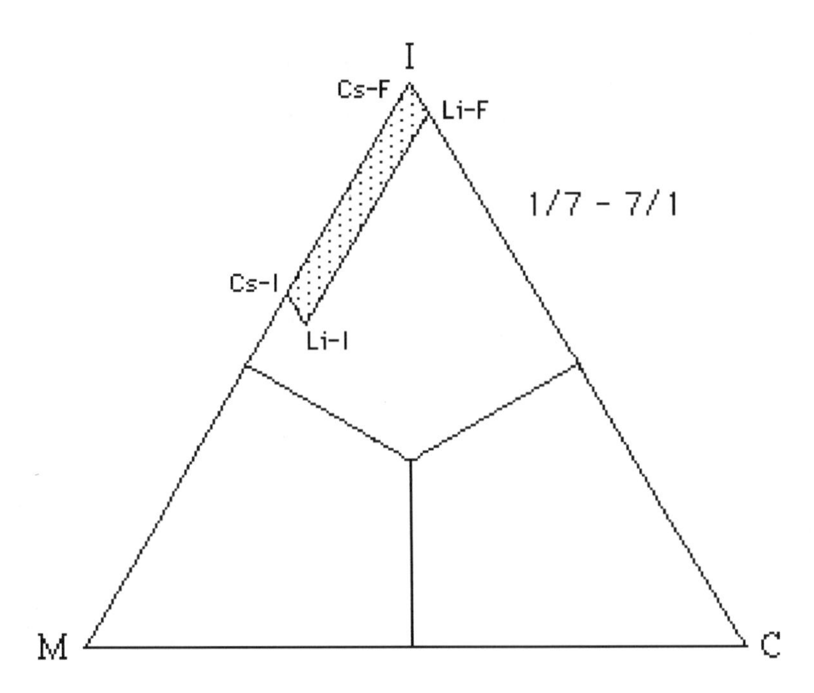

Traditionally this decrease in ionicity has been attributed to increasing covalent character due to the greater polarizing ability of the small Li^+ cation versus that of the larger Cs^+ ion and to the greater polarizability of the soft I^- anion versus the hard F^- ion (6). But, in fact, because of the small variation in *FEN* values for the alkali metals, which determines the width of the parallelogram, the covalent character of these bonds varies by only 5% on passing from the Cs bonds to the Li bonds, whereas its much greater length is determined by the much greater variation in the *FEN* values of the halogens. Thus, as correctly inferred by the brilliant work of Zintl in the 1930s, most of the change in ionic character is actually coupled, not with a change in covalency, but rather with a concomitant change in the metallicity of the bonds, which varies from a minimum of 0% for the fluoride bonds to a maximum of 38% for the iodide bonds – a trend which will continue throughout the remaining six triangles discussed below (7).

As shown in Section 3.4, the width of the parallelogram is determined by the electropositive branches of the most and least electropositive A component of the various compounds, whereas its length is determined by the electronegative branches of the most and least electronegative B component. As already mentioned, the data points for all other bonds in the table lie within the resulting parallelogram.

5.4 Group *1/7* - Group *6/2* Binaries

The net stoichiometries of most of the known binary compounds formed between the alkali metals and the chalcogens or Group *6/2* elements are summarized in the following table (4, 8):

B \ A	Li	Na	K	Rb	Cs
O	Li_2O Li_2O_2	Na_2O Na_2O_2 NaO_2 NaO_3	K_2O K_2O_2 KO_2 KO_3	Rb_6O Rb_9O_2 Rb_2O Rb_2O_2 RbO_2 RbO_3	Cs_4O Cs_7O etc Cs_2O Cs_2O_2 CsO_2 CsO_3
S	Li_2S	Na_2S Na_2S_b b = 2-5	K_2S K_2S_b b = 2-6	Rb_2S Rb_2S_b b = 2-6	Cs_2S Cs_2S_b b = 2-6
Se	Li_2Se	Na_2Se Na_2Se_b b = 2-6	K_2Se K_2Se_b b = 2-6	Rb_2Se Rb_2Se_b b = 2-4	Cs_2Se
Te	Li_2Te	Na_2Te Na_2Te_b b = 2-5	K_2Te K_2Te_b b = 2-3	Rb_2Te Rb_2Te_b b = 2-3	Cs_2Te

From relation 6 the stoichiometry for the simple homodesmic binaries is predicted to be A_2B. Thus for homodesmic binaries: $b/a = e_A/v_B = 1/2$, for polyanionic heterodesmic binaries: $b/a > 1/2$, and for polycationic heterodesmic binaries: $b/a < 1/2$. Again, all of the resulting A-B bonds may be illustrated using highly stable, simple homodesmic binaries (e.g. Li_2O, Na_2S, K_2Se, etc.). It is also apparent that polyanionic compounds are far more numerous than was the case for the halides. Examples include $A(B_2)$, $A_2(B_2)$ and $A(B_3)$ phases containing the superoxide (O_2^-), peroxide (O_2^{2-}), and the ozonide (O_3^-) anions, respectively, as well many $A_2(B_b)$ phases

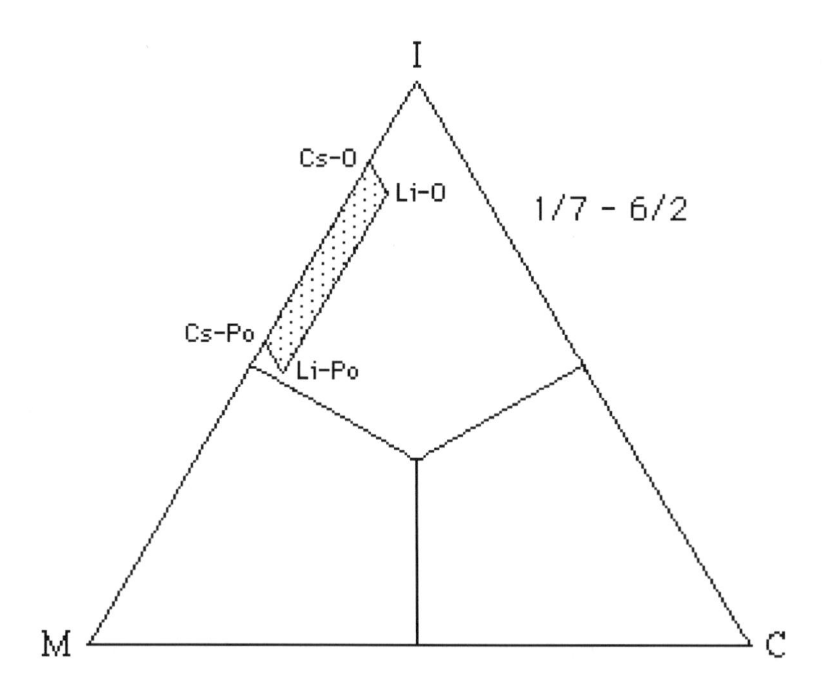

containing discrete polysulfide, polyselenide and polytelluride chain and ring anions (S_3^{2-}, S_4^{2-}, Se_2^{2-}, Se_8^{2-}, Te_5^{2-}, etc.) of varying sizes. In the case of Rb and Cs, polycationic binaries such as $(Rb_6)O$ and $(Cs_4)O$ are also known.

As was the case with the halogens, the above bond-type triangle once again places all of the resulting A-B bonds within a parallelogram located along the upper left-hand edge of the triangle within the predominantly ionic region, with the ionic character ranging from a maximum of 85% for the Cs-O bond to a minimum of 49% for the Li-Po bond. As before, the covalency variation remains at 5%, whereas the metallicity varies from a minimum of 15% for the oxide bonds to a maximum of 46% for the polonide bonds. Note that there is a substantial overlap between the parallelogram for the halides bonds and that for the chalcide bonds reflecting the fact that the *FEN* values of O and S overlap with those of the halogens.

5.5 Group *1/7* - Group *5/3* Binaries

The net stoichiometries of most of the known binary compounds formed between the alkali metals of Group *1/7* and the pnictogens or Group *5/3* elements are summarized in the following table (4, 7-9):

B \ A	Li	Na	K	Rb	Cs
N	Li_3N LiN_3	Na_3N NaN_3	KN_3	RbN_3	CsN_3
P	Li_3P LiP Li_3P_7 LiP_5 LiP_7 LiP_{15}	Na_3P NaP Na_3P_7 Na_3P_{11} NaP_7 NaP_{15}	K_3P KP K_4P_6 K_3P_7 K_3P_{11} KP_{15}	Rb_4P_6 Rb_3P_7 Rb_3P_{11} RbP_{11} RbP_{15}	Cs_4P_6 Cs_3P_7 Cs_3P_{11} CsP_7 CsP_{11} CsP_{15}
As	Li_3As $LiAs$	Na_3As $NaAs$ Na_3As_7 $NaAs_5$	K_3As K_5As_4 KAs KAs_2	Rb_3As	Cs_3As
Sb	Li_3Sb	Na_3Sb $NaSb$	K_3Sb K_3Sb_4 KSb KSb_2	Rb_3Sb Rb_5Sb_2 Rb_5Sb_4 $RbSb$ $RbSb_2$ Rb_3Sb_7	Cs_3Sb Cs_5Sb_2 Cs_2Sb Cs_3Sb_2 Cs_5Sb_4 $CsSb$ Cs_3Sb_7
Bi	Li_3Bi $LiBi$	Na_3Bi $NaBi$	K_3Bi K_3Bi_2 K_5Bi_4 KBi_2	Rb_3Bi Rb_3Bi_2 Rb_5Bi_4 $RbBi_2$	Cs_3Bi Cs_3Bi_2 Cs_5Bi_4 $CsBi_2$

From relation 6 the stoichiometry for the simple homodesmic binaries is predicted to be A_3B. Thus for homodesmic binaries: $b/a = e_A/v_B = 1/3$, for polyanionic heterodesmic binaries: $b/a > 1/3$, and for polycationic heterodesmic binaries: $b/a < 1/3$. As may be seen, in the case of the nitrides only Li and Na form stable simple homodesmic binaries and the remaining metal-nitrogen bonds must instead be illustrated using polyanionic compounds of the composition $A(B_3)$, which contain the discrete linear azide or polytrinitride anion (N_3^-). On the other hand, in the case of the heavy pnictogens all of the simple binary compounds are apparently known, with the exceptions of Rb_3P and Cs_3P, and there is also a considerable range of known polyanionic species corresponding not only to discrete chains (P_2^{4-}), and polycycles (P_7^{3-}) but also to infinite chain (P^-) and layer (P_6^{4-}) structures.

Though, in the bond-type triangle at the top of the following page, the parallelogram containing the resulting A-B bonds is also located along the

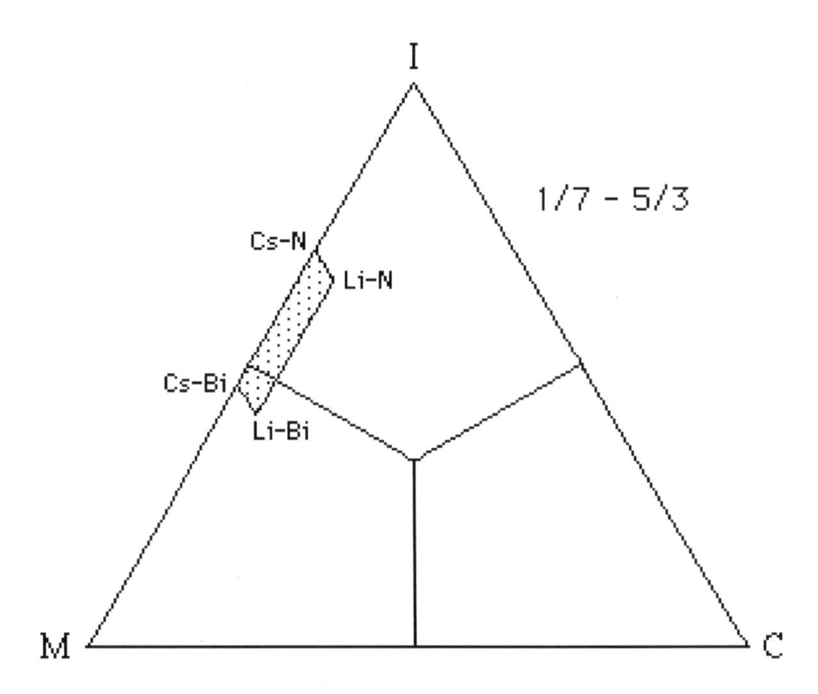

left-hand edge of the bond-type triangle, it now lies partially within the predominantly ionic region and partially within the predominantly metallic region, with the ionic character varying between 70% for the Cs-N bond and 39% for the Li-Bi bond, and the metallic character simultaneously varying between 30% for the nitride bonds and 54% for the bismuthide bonds. Again there is overlap between this parallelogram and the previous one for the chalcogens due to the overlap of the *FEN* value of N with those of the chalcogens.

5.6 Group *1/7* - Group *4/4* Binaries

The net stoichiometries of most of the known binary compounds formed between the alkali metals of Group *1/7* and the carbon or Group *4/4* elements are summarized in the table at the top of the following page (4, 7-9).

From relation 6 the stoichiometry for the simple homodesmic binaries is predicted to be A_4B. Thus for homodesmic binaries: $b/a = e_A/v_B = 1/4$, for polyanionic heterodesmic binaries: $b/a > 1/4$, and for polycationic hetero-

B \ A	Li	Na	K	Rb	Cs
C	Li_2C_2 LiC_b Li_aC_{60}	Na_2C_2 NaC_b Na_aC_{60}	K_2C_2 KC_b K_aC_{60}	Rb_2C_2 RbC_b Rb_aC_{60}	Cs_2C_2 CsC_b Cs_aC_{60}
Si	$Li_{22}Si_5$ Li_4Si Li_7Si_2 $Li_{10}Si_3$ Li_2S $Li_{13}S_7$	Na_4Si NaSi $NaSi_2$ Na_8Si_{46}	KSi K_8Si_{46} KSi_6	RbSi $RbSi_6$ $RbSi_8$	CsSi $CsSi_8$
Ge	$Li_{22}Ge_5$ $Li_{15}Ge_4$ Li_7Ge_2 Li_3Ge Li_9Ge_4 LiGe	NaGe	KGe KGe_4 K_8Ge_{46}	RbGe $RbGe_4$	CsGe $CsGe_4$
Sn	$Li_{22}Sn_5$ Li_4Sn Li_7Sn_2 Li_5Sn_2 Li_2Sn LiSn $LiSn_2$ Li_2Sn_5	$Na_{15}Sn_4$ Na_2Sn Na_3Sn Na_4Sn_3 NaSn $NaSn_2$ etc	K_2Sn KSn KSn_2 KSn_4 K_8Sn_{46}	RbSn	CsSn
Pb	$Li_{22}Pb_5$ Li_7Pb_2 Li_3Pb Li_8Pb_3 LiPb	Na_5Pb $Na_{15}Pb_4$ $Na_{13}Pb_5$ Na_5Pb_2 Na_9Pb_4 NaPb	K_2Pb KPb K_2Pb_3 K_4Pb_9 KPb_2 KPb_4	RbPb	CsPb

desmic binaries: *b/a < 1/4*. As may be seen, in the case of the carbides there are no simple homodesmic binaries containing only A-B bonds, and one must instead resort to polyanionic compounds of the compositions $A_2(C_2)$, $A(C)_b$ and $A_a(C_{60})$. The first of these contains the discrete polydicarbide or so-called acetylide anion C_2^{2-}; the second corresponds to the intercalation compounds of the alkali metals with graphite (where b = 6, 8, 10, 16, 18, 24, 32, 36, 48, 60 etc.), and the third to the intercalation compounds of the alkali metals with the so-called buckeyball form of carbon (where a = 1-3). Though a few simple binaries are known for the heavier Group *4/4* elements (e.g., Na_4Si, Li_4Sn, Li_4Si), one must resort once again to the rich variety of known polyanionic phases, which contain not only discrete anionic chains, rings, and polycycles, but also infinitely extended chain, layer, and framework substructures. In addition, we also encounter several potential polycationic compounds, such as $(Li_{22})Ge_5$ and $(Na_5)Pb$.

As shown on the bond-type triangle at the top of the following page, with the exception of the carbides, the resulting bond parallelogram now lies completely within the predominantly metallic region, though this by no means

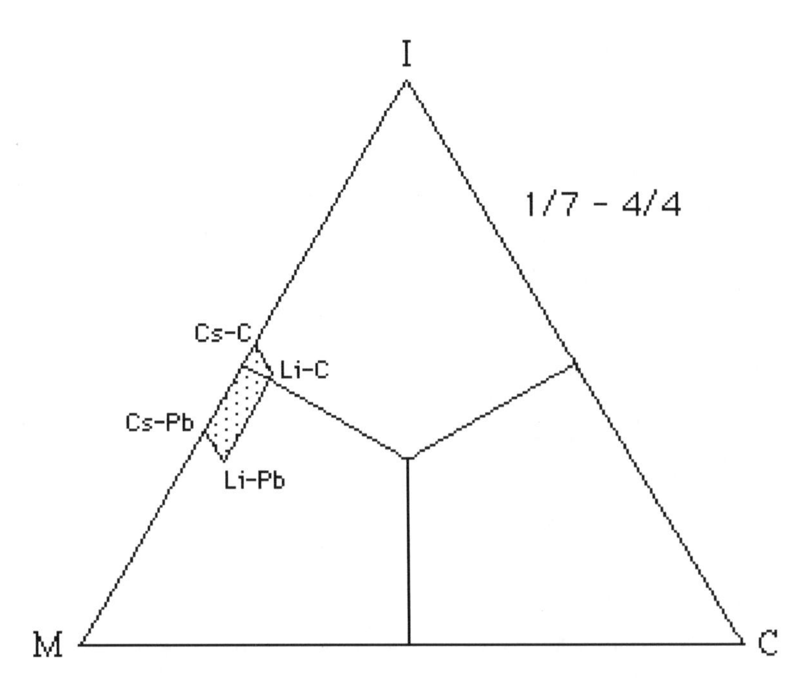

implies that the resulting compounds automatically possess the bulk properties of typical metals, such as ductility, malleability, high electronic conductivities and metallic sheen. Rather the majority are best described as poor semi-conductors. The ionic character of the resulting A-B bonds varies between 54% for the Cs-C bond and 33% for the Li-Pb bond, and the metallic character between 46% for the carbide bonds and 62% for the plumbide bonds. Once more there is overlap between this parallelogram and the previous one for the pnictides due to the overlap of the *FEN* value of C with those of the pnictogens.

5.7 Group *1/7* - Group *3/5* Binaries

The net stoichiometries of most of the known binary compounds formed between the alkali metals of Group *1/7* and the boron or Group *3/5* elements are summarized in the table at the top of the following page (4, 7-9).

From relation 6 the stoichiometry for the simple homodesmic binaries is predicted to be A_5B. Thus for homodesmic binaries: $b/a = e_A/v_B = 1/5$, for

B \ A	Li	Na	K	Rb	Cs
B	LiB_4 LiB_6	NaB_6	KB_6	RbB_6	CsB_6
Al	Li_3Al Li_9Al_4 $LiAl$ Li_2Al_3				
Ga	Li_2Ga $LiGa$	Na_5Ga_8 $NaGa_4$	K_5Ga_8 KGa_4	Rb_5Ga_8 $RbGa_4$	Cs_5Ga_8 $CsGa_4$
In	$Li_{20}In$ Li_9In Li_9In_2 Li_4In Li_3In Li_9In_4 Li_3In_2 $LiIn$	Na_2In $NaIn$ Na_5In_8	K_5In_8 KIn_4	Rb_5In_8 $RbIn_4$	Cs_5In_8 $CsIn_4$
Tl	Li_4Tl Li_3Tl Li_5Tl_2 Li_2Tl $LiTl$	Na_6Tl Na_2Tl $NaTl$ $NaTl_2$	KTl K_4Tl_5 K_5Tl_8	Rb_4Tl_5 $RbTl_2$ $RbTl_3$	Cs_5Tl_7 Cs_4Tl_7 $CsTl_3$

polyanionic heterodesmic binaries: *b/a > 1/5*, and for polycationic heterodesmic binaries: *b/a < 1/5*. No examples of simple homodesmic binaries are known, though, both polyanionic (e.g, $Na(B_6)$, $K(Ga_4)$, $Rb_5(In_8)$, $Cs_4(Tl_7)$ etc.) and polycationic species (e.g., $(Li_{20})In$, $(Na_6)Tl$, etc.) are common.

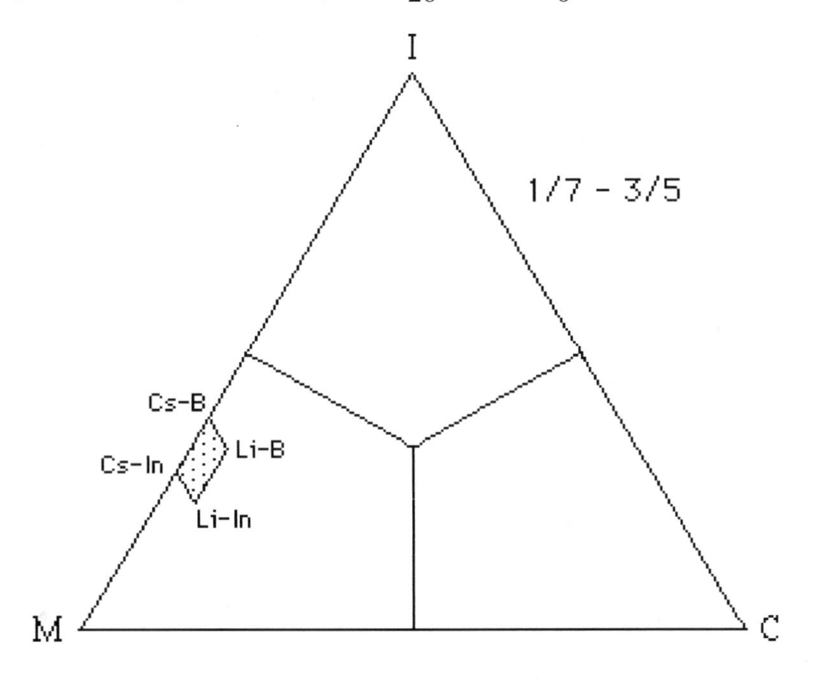

As shown at the bottom of the previous page, the corresponding bond parallelogram now lies completely within the predominantly metallic region, though again most of the resulting compounds do not possess the full range of typical bulk properties normally associated with true metals and are instead best described as semiconductors. The ionic character of the resulting A-B bonds varies between 38% for the Cs-B bond and 23% for the Li-In bond, whereas their metallic character varies between 62% for the boride bonds and 72% for the indide bonds. Here for the first time there is no overlap with the previous parallelogram. Note also that this particular parallelogram is unusual short. This is due both to the fact that the range of *FEN* values within a given group of the periodic table becomes progressively shorter as we move from the right to the left of the periodic table and to the fact that the range for Group *3/5* is unusually compressed and irregular due to the strong effects of both the transition-block and the inner-transition block insertions.

5.8 Group *1/7* - Group *2/6* Binaries

The net stoichiometries of most of the known binary compounds formed between the alkali metals of Group *1/7* and both the alkaline earth metals or Group *2/6* elements and the zinc or Group *2/6'* subgroup elements (i.e. Zn, Cd, and Hg) are summarized in the two tables on the following page (9).

From relation 6 the stoichiometry for the simple homodesmic binaries is predicted to be A_6B. Thus for homodesmic binaries: $b/a = e_A/v_B = 1/6$, for polyanionic heterodesmic binaries: $b/a > 1/6$, and for polycationic heterodesmic binaries $b/a < 1/6$. The resulting compounds are normally viewed as true intermetallic alloys which possess the full range of typical bulk properties normally associated with metals and are seldom if ever discussed in inorganic textbooks, thereby forcing one to have recourse to the metallurgical phase-diagram literature instead. As may be seen, this reveals that, in sharp contrast with the combinations with the Zn subgroup (and especially with Hg), most of the possible combinations with the alkaline earth metals apparently do not

B \ A	Li	Na	K	Rb	Cs
Be					
Mg					
Ca	Li_2Ca				
Sr	$Li_{23}Sr_6$ $LiSr_8$ Li_2Sr_3				
Ba	Li_4Ba	Na_4Ba			

B \ A	Li	Na	K	Rb	Cs
Zn	$LiZn$ Li_2Zn_3 $LiZn_2$ Li_2Zn_5 $LiZn_4$	$NaZn_{13}$	KZn_{13}	$RbZn_{13}$	
Cd	Li_3Cd $LiCd_3$	$NaCd_2$ Na_2Cd_{11}	KCd_{13}	$RbCd_{13}$	$CsCd_{13}$
Hg	Li_6Hg Li_3Hg Li_2Hg $LiHg$ $LiHg_2$ $LiHg_3$	Na_3Hg Na_5Hg_2 Na_3Hg_2 $NaHg$ Na_7Hg_8 $NaHg_2$ $NaHg_4$	KHg K_5Hg_7 KHg_2 $KHg_{2.7}$ KHg_4 KHg_3 KHg_9 KHg_{11}	Rb_7Hg_8 Rb_3Hg_4 $RbHg_2$ Rb_2Hg_5 Rb_2Hg_7 Rb_2Hg_9 $RbHg_6$ $RbHg_{11}$	$CsHg$ Cs_5Hg_4 $CsHg_2$ $CsHg_4$ $CsHg_6$ $CsHg_{12}$

form intermetallic compounds In those cases where compounds are formed, they have unusual stoichiometries (e.g., Li_3Cd, $Cs(Hg_4)$, $K(Zn_{13})$, etc.) which cannot be rationalized using simple valence rules and which are either polyanionic or contain a mixture of A-B, A-A and B-B bonds.

As shown on the following bond-type triangle, the parallelogram containing the few resulting A-B bonds lies completely within the predominantly metallic region, where, for purposes of our brief survey, we have not bothered to separate out the 2/6' subgroup combinations, though this easily done if so desired. The ionic character of the resulting A-B bonds varies between 25% for the Cs-Hg bond and 5% for the Li-Ba bond, while their metallic character varies between 75% for the mercuride bonds and 90% for the baride bonds. Note that not only is there no overlap with the previous parallelogram, there is actually a gap due to the 0.03 *FEN* difference separating the most electronegative Group *2/6'* element (Hg) from the least electronegative Group *3/5* element (In).

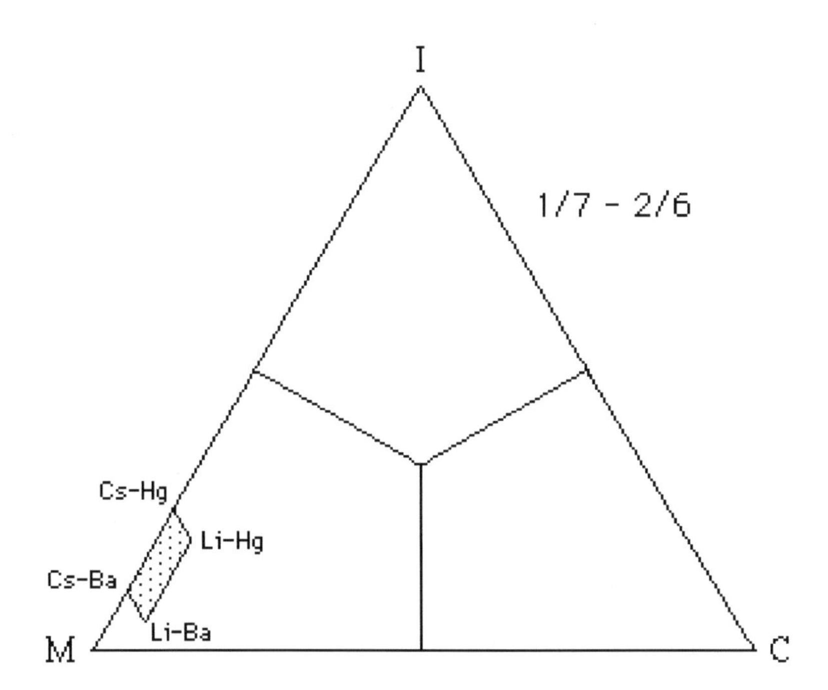

5.9 Group *1/7* - Group *1/7* Binaries

The net stoichiometries of most of the known binary compounds formed among the alkali metals of Group *1/7* are summarized in the following table (9):

B \ A	Li	Na	K	Rb	Cs
Li					
Na			KNa_2		$CsNa_2$
K					CsK_2
Rb					
Cs					

From relation 6 the stoichiometry for the simple homodesmic binaries is predicted to be A_7B. Thus for homodesmic binaries: $b/a = e_A/v_B = 1/7$, for polyanionic heterodesmic binaries: $b/a > 1/7$, and for polycationic heterodesmic binaries: $b/a < 1/7$. As with the case of the Group *1/7* - Group *2/6* combinations, most inorganic textbooks report nothing concerning intermetallic compounds among the various alkali metals and consultation of the metallurgical phase-diagram literature once more reveals that the vast majority of possible binary combinations apparently form no compounds at all. Only three very unstable compounds have been reported, all of them corresponding to a probable stoichiometry of AB_2, and which, in all probability, are either polyanionic or contain a mixture of A-B, A-A and B-B bonds.

In the case of the corresponding bond-type triangle, the usual parallelogram has been truncated to a small triangle due to its intersection with the metallic corner of the triangle. This automatically eliminates redundant but inverted bond combinations. Thus Cs-Li is retained but the redundant Li-Cs, which inverts our established convention of always writing the more electropositive component first, is automatically eliminated. The ionic character of

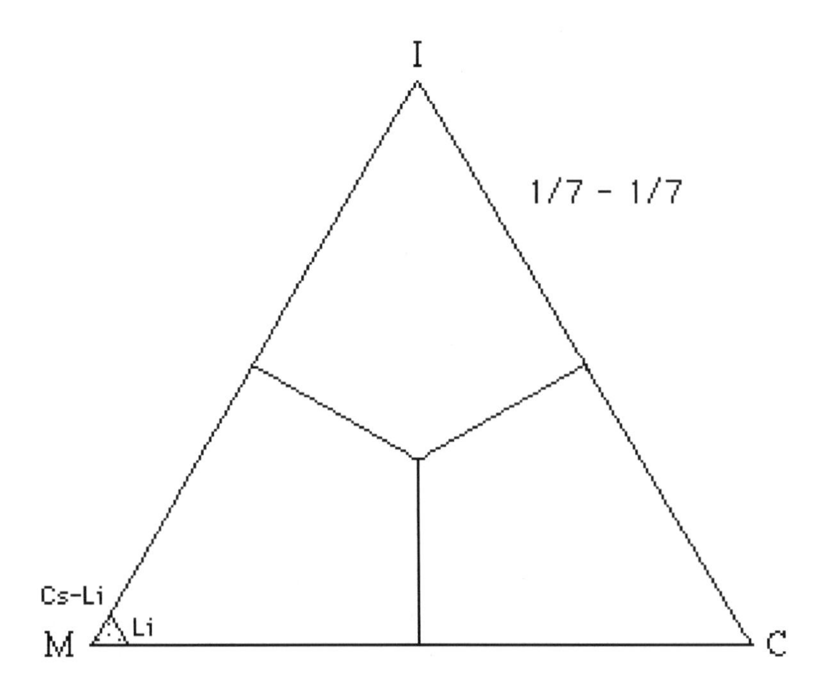

the resulting A-B bonds ranges from a maximum of 5% for the apparently hypothetical Cs-Li bond to a minimum of 0% for all of the self A-A bonds lying along the bottom edge of the triangle. Likewise the metallic character varies from a maximum of 100% by definition for the Cs-Cs bond to a minimum of 95% for the Li combinations, whereas the covalent character of the simple substances lying along the base varies from 0% by definition for the Cs-Cs bond to 5% for the Li-Li bond. Because of a 0.05 *FEN* difference separating the most electronegative Group *1/7* element (Li) from the least electronegative Group *2/6* element (Ba), there is again a gap separating this triangle from the previous parallelogram.

5.10 References and Notes

1. For a brief discussion of an earlier version of this table, see W. B. Jensen, "Richard Abegg and the Periodic Table" in E. Scerri, G. Restrepo, Eds., *Mendeleev to Organesson: A Multidisciplinary Perspective on the Periodic Table*, Oxford University Press, New York, NY, 2018, pp. 245-265. A more detailed discussion will appear in the forthcoming book, W. B. Jensen, *From Triads to Aufbau: Lectures on the History and*

Nature of the Periodic Law and Table.

2. A. Savin, R. Nesper, S. Wengert, T. Fässler, "ELF: The Electron Localization Function," *Angew. Chem. Int. Ed. Engl.*, **1997**, *36*, 1808-1832.

3. P. G. Blake, D. W. Clack, "Cesium Neonide: Molecule or Thermochemical Exercise?," *J. Chem. Educ.*, **1982**, *59*, 637-639.

4. N. Wiberg, *Inorganic Chemistry*, Academic Press: New York, NY, 1995, pp. 421-422; 686-689; 794-797; 1104-1107; 1107-1111

5. Some structural information may be imposed on the net stoichiometric formulas using parentheses to indicate the existence of A-A or B-B bonded substructures. When the substructure is finite the corresponding stoichiometric coefficient in placed within the parentheses as in $(Hg_2)Cl_2$, which contains the discrete Hg_2^{2+} cation, or $Ca(C_2)$, which contains the discrete C_2^{2-} anion. When it is infinitely extended, the coefficient is placed outside the brackets as in $Ca(Si)_2$, which contains an infinitely extended Si^- layer anion, or (C)F, which contains infinitely extended C layers. For details and further elaborations see W. B. Jensen, "Crystal Coordination Formulas: A Flexible Notation for the Interpretation of Solid-State Structures," in D. G. Pettifor, Ed., *The Structures of Binary Compounds*, North-Holland: Amsterdam, 1989, pp. 105-146.

6. B. E. Douglas, D. H. McDaniel, J. J. Alexander, *Concepts and Models in Inorganic Chemistry*, 3rd ed., Wiley: New York, NY, 1994, pp. 200-203.

7. H. Schäfer, B. Eisenmann, W. Müller, "Zintl Phases: Transitions Between Metallic and Ionic Bonding," *Angew. Chem. Int. Ed. Engl.*, **1973**, *12*, 694-712.

8. J. C. Bailar et al., Eds., *Comprehensive Inorganic Chemistry*, Vol. 1, Pergamon Press: Oxford, 1973, pp. 702-703, 1205, 1272-1278; Vol. 2, p. 280.

9. B. Predel, Ed., *Phase Equilibria, Crystallographic and Thermodynamic Data of Binary Alloys,* Landolt-Börnstein, Vols. 5a-5i, Springer: Berlin, 1991-1998.

VI
Bonding in the Binary Compounds of Group *2/6*

6.1 Zintl Phases

The trends for the bonds formed between the alkaline earth or Group *2/6* elements and the rest of the main-block elements are very similar to those already observed for the alkali metals or Group *1/7* elements and they have been accorded a separate lecture solely for reasons of length. Once again, we encounter a progressive transition from predominantly ionic bonding to predominantly metallic bonding, though the variation in both covalent character and the width of the resulting bond parallelograms is greater (15% versus 5%). Indeed the binary compounds of both the Group *1/7* and Group *2/6* elements are often grouped together under the rubric of "Zintl Phases" in honor of the German chemist, Eduard Zintl, who first studied the ionic-metallic bond transition in detail (1). It should be noted that this similarity has nothing to do with a fundamental discontinuity between so-called s-block versus p-block elements, as the elements in both of these blocks make full use of both their filled and empty s- and p-orbitals in their bonding interactions. Rather it is related to the low electronegativity values and the relatively large sizes of both the Group *1/7* and Group *2/6* elements.

In what follows, we have not bothered to explicitly separate out the Zn or Group *2/6'* elements in our bond triangles. We have, nevertheless, provided separate tables for their binary compounds. Though often incorrectly classified as transition-block elements, there is no doubt that Zn, Cd, and Hg are main-block elements and that a fundamental bifurcation in Group *2/6* occurs after Mg leading to both pre- and post-transition branches (2). Since the *FEN* value for Hg (0.25), which is the largest for the post-transition branch, is very similar to that of Be (0.24), the individual bond parallelograms for Groups *2/6* and *2/6'* essentially overlap. Indeed, 19th- and early 20th-century chemistry texts frequently classified Be and Mg with the Zn branch rather than with the Ca branch (due, as we now recognize, to their similar *EN* values) and often used the name "alkaline-earth metals" to denote only the members of the latter (i.e. Ca-Ra).

Zintl also provided the experimental data for the formulation of a rule governing the structures of the binary compounds of the Group *1/7* and Group *2/6* elements. It is based on a specific application of the isoelectronic principle known the Klemm-Busmann principle (3, 4). It states that, if one assumes that the A-B bond in the binary compound is predominately ionic, such that all of the valence electrons on A have been transferred to the more electronegative component B, then B will have a structure similar to that of a neutral simple substance containing the same number of valence electrons and valence vacancies. The resulting possibilities are summarized in the following table:

Ionic Charge					
0	Group *4/4*	Group *5/3*	Group *6/2*	Group *7/1*	Group *8/0*
1-	Group $(3/7)^{1-}$	Group $(4/4)^{1-}$	Group $(5/3)^{1-}$	Group $(6/2)^{1-}$	Group $(7/1)^{1-}$
2-		Group $(3/7)^{2-}$	Group $(4/4)^{2-}$	Group $(5/3)^{2-}$	Group $(6/2)^{2-}$
3-			Group $(3/7)^{3-}$	Group $(4/4)^{3-}$	Group $(5/3)^{3-}$
4-				Group $(3/7)^{4-}$	Group $(4/4)^{4-}$

Thus, for example, in the compound NaTl, the Group *3/5* anion has a formal charge of 1-, thus making it isoelectronic with a neutral Group *4/4* element, and thus giving the anion a structure similar to the diamond structure of carbon. In the compound ZnP_2, each of the two Group *5/3* anions has a formal charge of 1-, thus making them isoelectronic with a neutral Group *6/2* element, and giving the anion an infinite chain structure like that found in the gray form of selenium. In CaO, the Group *6/2* anion has a formal charge of 2-, making it isoelectronic with a neutral Group *8/0* element like neon, and thus giving a discrete anion with a complete octet. In summary, only compounds having anions with charges that make them isoelectronic with the noble gases in Group *8/0* (i.e. fall into the last column of the table) will form simple binaries. All other cases in the table will form polyanionic compounds containing some form of B-B bonding in addition to the A-B bonds.

6.2 Group *2/6* -Group *7/1* Binaries

The net stoichiometries of most of the known binary compounds formed between the Group *2/6* elements and the halogens or Group *7/1* elements are summarized in the two tables on the following page (5, 7).

The stoichiometry for homodesmic or simple binaries is predicted to be AB_2. Thus for homodesmic binaries: $b/a = e_A/v_B = 2/1$, for heterodesmic polyanionic binaries: $b/a > 2/1$, and for heterodesmic polycationic binaries: $b/a < 2/1$. As may be seen, all of the resulting A-B bonds may be illustrated using stable, simple, binary compounds. In contrast to Group *1/7*, there are no anhydrous polyiodides – a fact which appears to be a consequence of the principle that large anions are best stabilized using counter cations of similar size and equal, but opposite, charge (6). In the case of mercury one also notes the appearance of stable polycationic compounds containing the so-called mecurous or Hg_2^{2+} ion. The analogous polycationic species Zn_2^{2+} and Cd_2^{2+} have been detected spectroscopically in reduced molten chloride melts but have not been isolated within pure binary compounds.

In keeping with the standard textbook characterization of these bonds as ionic, we see in the bond-type triangle on page 118 that, with the exception

B \ A	Be	Mg	Ca	Sr	Ba
F	BeF_2	MgF_2	CaF_2	SrF_2	BaF_2
Cl	$BeCl_2$	$MgCl_2$	$CaCl_2$	$SrCl_2$	$BaCl_2$
Br	$BeBr_2$	$MgBr_2$	$CaBr_2$	$SrBr_2$	$BaBr_2$
I	BeI_2	MgI_2	CaI_2	SrI_2	BaI_2

B \ A	Zn	Cd	Hg
F	ZnF_2	CdF_2	Hg_2F_2 HgF_2
Cl	Zn_2Cl_2 $ZnCl_2$	Cd_2Cl_2 $CdCl_2$	Hg_2Cl_2 $HgCl_2$
Br	$ZnBr_2$	$CdBr_2$	Hg_2Br_2 $HgBr_2$
I	ZnI_2	CdI_2	Hg_2I_2 HgI_2

of the borderline Hg-I bond, they are all located within a parallelogram lying within the predominantly ionic region, with their ionic character ranging from a maximum of 90% for the Ba-F bond to a minimum of 37% for the Hg-I bond, and their metallic character ranging from 0% for the fluoride bonds to 38% for the iodide bonds. The covalency varies from 10% for Ba bonds to 25% for Hg bonds. The parallelogram is wider than was the case for the Group *1/7* elements because the difference in the *FEN* values of Ba versus Hg is greater than that between Li and Cs.

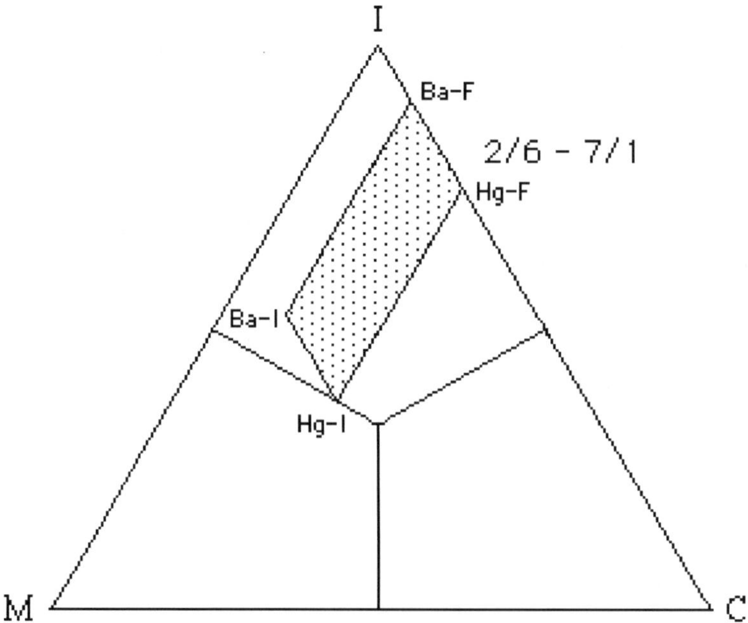

6.3 Group *2/6* - Group *6/2* Binaries

The net stoichiometries of most of the known binary compounds formed between the Group *2/6* elements and the chalcogens or Group *6/2* elements are summarized in the following two tables (1, 5, 7):

B \ A	Be	Mg	Ca	Sr	Ba
O	BeO	MgO MgO_2	CaO CaO_2 CaO_4 CaO_6	SrO SrO_2 SrO_4	BaO BaO_2 BaO_4 BaO_6
S	BeS	MgS	CaS	SrS SrS_2 SrS_3	BaS Ba_2S_3 BaS_2 BaS_3
Se	BeSe	MgSe $MgSe_2$	CaSe	SrSe Sr_2Se_3 $SrSe_2$ $SrSe_3$	BaSe $BaSe_2$ $BaSe_3$
Te	BeTe	MgTe $MgTe_2$	CaTe	SrTe Sr_2Te_3 $SrTe_2$	BaTe Ba_2Te_3 $BaTe_2$

B \ A	Zn	Cd	Hg
O	ZnO ZnO_2	CdO CdO_2	HgO
S	ZnS	CdS	HgS
Se	ZnSe	CdSe	HgSe
Te	ZnTe	CdTe	HgTe

The stoichiometry for homodesmic or simple binaries is predicted to be AB. Thus for homodesmic binaries: $b/a = e_A/v_B = 2/2 = 1/1$, for heterodesmic polyanionic binaries: $b/a > 1/1$, and for heterodesmic polycationic binaries: $b/a < 1/1$. Examples of all of these bonds are known in the form of simple binaries. Polyanionic peroxides $AO_2 = A(O_2)$ are also known for all of the elements except Be and Hg and examples of both superoxides $AO_4 = A(O_2)_2$ and ozonides $AO_6 = A(O_3)_2$ have been prepared for Ca, Sr, and Ba. Though some polyanionic compounds of the form $A(B_b)$ are also known for the heavier chalcogens, they are surprisingly limited compared with the heavier polychalcides of the alkali metals. In particular, the paucity of known examples for Ca is apparently due to the fact that the corresponding binary phases diagrams have never been determined (7). Also of note is the absence of any mercury (I) chalcide compounds – the black oxide, Hg_2O, often assumed in the thermal decomposition of HgO, apparently being a mixture of Hg and HgO (8).

In the bond-type triangle on the following page, the parallelogram for these bonds, which overlaps with that for the halide bonds, lies largely within the predominantly ionic region with only a small protrusion into the predominantly metallic region. The ionic character ranges from a maximum

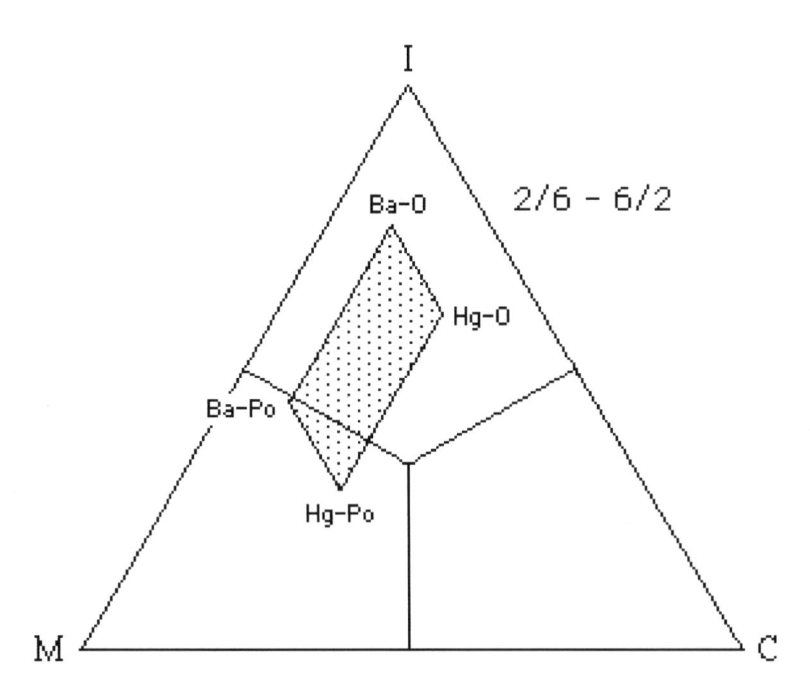

of 75% for the Ba-O bond to a minimum of 29% for the Hg-Po bond. As before, the covalency varies from 10% for Ba bonds to 25% for Hg bonds, whereas the metallicity varies from a minimum of 15% for the oxide bonds to a maximum of 46% for the polonide bonds

6.4 Group *2/6* - Group *5/3* Binaries

The net stoichiometries of most of the known binary compounds formed between the Group *2/6* elements and the pnictogens or Group *5/3* elements are summarized in the two tables on the following page (1, 7).

The stoichiometry for homodesmic or simple binaries is predicted to be A_3B_2. Thus for homodesmic binaries: $b/a = e_A/v_B = 2/3$, for heterodesmic polyanionic binaries: $b/a > 2/3$, and for heterodesmic polycationic binaries: $b/a < 2/3$. In the case of the alkaline earth metals, examples of most of these bonds are known either in the form of simple binaries or in the form of a variety of polyanionic compounds – the sole exception apparently being the Be-Bi bond. A number of polycationic compounds are also apparent, such as

B \ A	Be	Mg	Ca	Sr	Ba
N	Be_3N_2	Mg_3N_2 MgN_6	Ca_3N_2 $Ca_{11}N_8$ Ca_3N_4 CaN_6	Sr_2N SrN SrN_6	Ba_2N Ba_3N_2 Ba_3N_4 BaN_2 BaN_6
P	Be_3P_2 BeP_2	Mg_3P_2 MgP_4	Ca_3P_2 CaP CaP_5	Sr_2P Sr_3P_2 SrP Sr_3P_4 Sr_3P_{14} SrP_3	Ba_2P Ba_3P_2 Ba_4P_3 BaP Ba_4P_5 BaP_2 BaP_3
As	Be_2As	Mg_3As_2 $MgAs_4$	Ca_2As Ca_3As_2 $CaAs$ $CaAs_4$	Sr_2As Sr_5As_3 Sr_3As_2 $SrAs$	Ba_5As_3 Ba_3As_2 $BaAs_2$
Sb	$Be_{13}Sb$ Be_3Sb_2 $BeSb_2$	Mg_3Sb_2	Ca_2Sb Ca_5Sb_3 Ca_3Sb_2 $CaSb_2$ $Ca_{11}Sb_{10}$	Sr_2Sb Sr_5Sb_3 Sr_3Sb_2 $SrSb$ Sr_2Sb_3 $SrSb_3$	Ba_2Sb Ba_5Sb_3 Ba_3Sb_2 Ba_5Bi_4 $BaSb_3$
Bi		Mg_3Bi_2	Ca_2Bi Ca_7Bi_4 Ca_5Bi_3 Ca_3Bi_2 $CaBi$ $CaBi_3$	Sr_2Bi Sr_3Bi_2 $SrBi$ $SrBi_3$	Ba_2Bi Ba_5Bi_3 Ba_3Bi_2 $BaBi$ $BaBi_3$

B \ A	Zn	Cd	Hg
N	Zn_3N_2 ZnN_6	Cd_3N_2 CdN_6	Hg_3N Hg_3N_2 HgN_3
P	Zn_3P_2 ZnP_2 ZnP_4	Cd_3P_2 Cd_6P_7 Cd_7P_{10} CdP_2 CdP_4	
As	Zn_3As_2 $ZnAs_2$	Cd_3As_2 $CdAs_2$	
Sb	Zn_3Sb_2 Zn_4Sb_3 $ZnSb$	Cd_3Sb_2 Cd_4Sb_3 $CdSb$ $CdSb_3$	Hg_3Sb_2
Bi			

those having the stoichiometry $(A_2)B$. In the case of the Zn subgroup, however, a far greater number of potential binary combinations are missing, especially in the case of Hg. The compound HgN_3 or mercurous azide is probably $(Hg_2)(N_3)_2$ and is thus a good example of a complex binary containing not only A-B bonds, but both A-A and B-B bonds as well.

The resulting bond parallelogram, which overlaps with that for the chalcide bonds, lies roughly half within the predominantly ionic region and half within the predominantly metallic region, with the ionic character ranging from a maximum of 60% for the Ba-N bond to a minimum of 21% for the apparently hypothetical Hg-Bi bond. As before, the covalency varies from 15% for the Ba bonds to 25% for the Hg bonds, whereas the metallicity varies from a minimum of 30% for the nitride bonds to a maximum of 54% for the bismuthide bonds. As was the case with the Group *1/7* elements, the bonds lying in the predominantly metallic region form binary compounds having properties corresponding to semiconductors rather than to typical bulk metals.

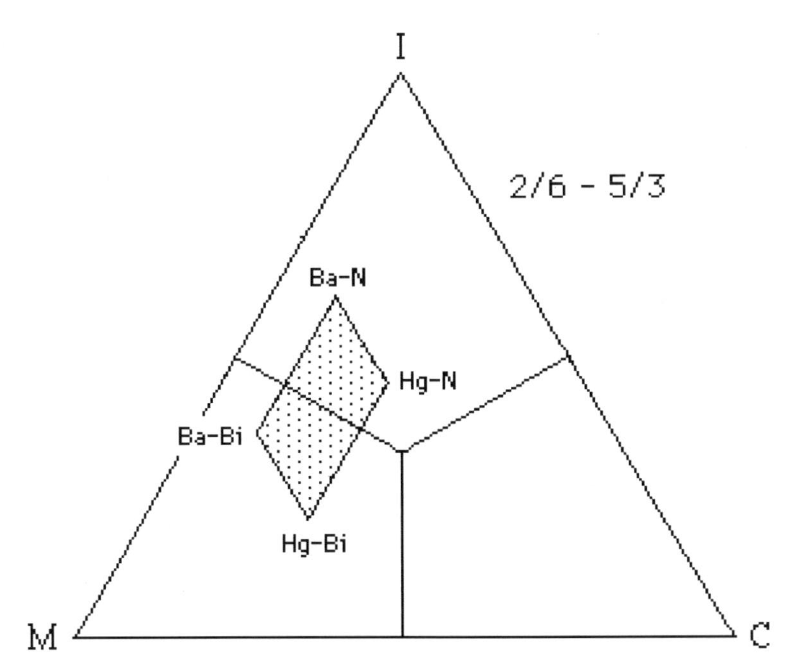

6.5 Group *2/6* - Group *4/4* Binaries

The net stoichiometries of most of the known binary compounds formed between the Group *6/2* elements and the carbon or Group *4/4* elements are summarized in the following two tables (1, 7):

B \ A	Be	Mg	Ca	Sr	Ba
C	$\underline{Be_2C}$	Mg_2C_3 MgC_2	CaC_2	SrC_2	BaC_2 BaC_6 BaC_8
Si		$\underline{Mg_2Si}$	$\underline{Ca_2Si}$ Ca_5Si_3 $CaSi$ $CaSi_2$	$\underline{Sr_2Si}$ $SrSi$ $SrSi_2$	$\underline{Ba_2Si}$ Ba_5Si_3 $BaSi$ Ba_3Si_4 $BaSi_2$
Ge		$\underline{Mg_2Ge}$	$\underline{Ca_2Ge}$ $CaGe$ $CaGe_2$	$\underline{Sr_2Ge}$ Sr_4Ge_3 $SrGe$ $SrGe_2$	$\underline{Ba_2Ge}$ $BaGe$ $BaGe_2$
Sn		$\underline{Mg_2Sn}$	$\underline{Ca_2Sn}$ $CaSn$ $CaSn_3$	$\underline{Sr_2Sn}$ Sr_5Sn_3 $SrSn$ $SrSn_3$ $SrSn_4$	$\underline{Ba_2Sn}$ Ba_5Sn_3 $BaSn$
Pb		$\underline{Mg_2Pb}$	$\underline{Ca_2Pb}$ Ca_5Pb_3 $CaPb$ $CaPb_3$	$\underline{Sr_2Pb}$ Sr_5Pb_3 $SrPb$ Sr_2Pb_3 Sr_3Pb_5 $SrPb_3$	$\underline{Ba_2Pb}$ Ba_5Pb_3 $BaPb$ Ba_3Pb_5 $BaPb_3$

B \ A	Zn	Cd	Hg
C	ZnC_2		HgC_2
Si			
Ge			
Sn			
Pb			$HgPb_2$

The stoichiometry for homodesmic or simple binaries is predicted to be A_2B. Thus for homodesmic binaries: $b/a = e_A/v_B = 2/4 = 1/2$, for heterodesmic polyanionic binaries: $b/a > 1/2$, and for heterodesmic polycationic binaries: *b/a* < *1/2*. The most striking thing about these tables is the almost total absence of binary compounds other than the carbides for those Group *2/6* elements have *FEN* values greater than 0.22 (Be, Zn, Cd and Hg) and the relatively small number known for Mg (*FEN* = 0.19) as compared with those for the elements of the pre-transition or Ca branch (*FEN* = 0.10-0.13). About 38% of the possible combinations form simple binaries and another 30% polyanionic species of the composition $A(B_2)$ which contain simple B_2^{2-} polyanions. The species $Ba(C)_6$ and $Ba(C)_8$ are intercalation compounds of graphite. Also of note is the absence of any polycationic species.

The resulting bond parallelogram now lies completely within the predominantly metallic region, though the resulting binary compounds are once again best described as semiconductors rather than as true metals. The ionic character of the bonds varies between 44% for the Ba-C bond and 13% for the apparently hypothetical Hg-Pb bond, whereas their metallic character varies between 46% for the carbide bonds and 62% for the plumbide bonds and the covalency between 10% for the Ba bonds and 25% for the Hg bonds.

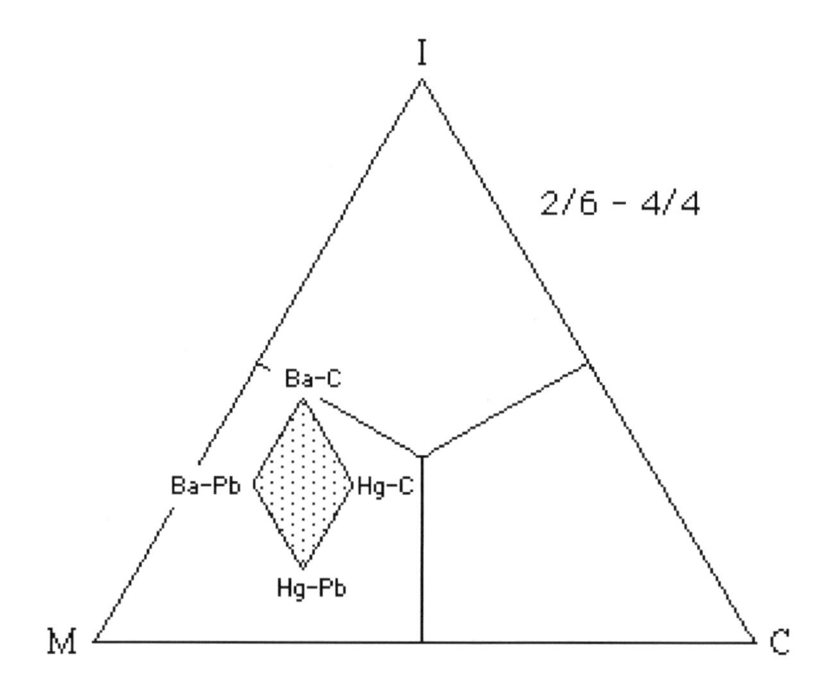

6.6 Group *2/6* - Group *3/5* Binaries

The net stoichiometries of most of the known binary compounds formed between the Group *6/2* elements and the boron or Group *3/5* elements are summarized in the following tables(1, 5):

B \ A	Be	Mg	Ca	Sr	Ba
B	Be_2B Be_2B_3 BeB_2 BeB_3 BeB_4 BeB_6 BeB_9 BeB_{12}	MgB_2 MgB_4 MgB_6	CaB_6	SrB_6	BaB_6
Al		Mg_2Al_3 $MgAl_3$ $MgAl_2$	$CaAl_2$ $CaAl_4$	$SrAl$ $SrAl_2$ $SrAl_4$	Ba_9Al Ba_3Ga_2 $BaAl$ $BaAl_2$ $BaAl_4$
Ga		Mg_5Ga_2 Mg_2Ga $MgGa$ $MgGa_2$ Mg_2Ga_5	Ca_3Ga Ca_5Ga_2 Ca_2Ga $CaGa$ $CaGa_2$ Ca_3Ga_8 $CaGa_4$	Sr_3Ga_2 $SrGa_2$ $SrGa_4$	$Ba_{10}Ga$ Ba_8Ga_7 $BaGa$ $BaGa_2$ $BaGa_4$
In		Mg_3In Mg_5In_2 Mg_2In $MgIn$ Mg_2In_5	Ca_3In $CaIn$ $CaIn_3$	Sr_3In Sr_3In_2 $SrIn$ Sr_2In_3 $SrIn_2$ $SrIn_3$	$Ba_{13}In$ Ba_3In Ba_2In $BaIn$ $BaIn_2$ $BaIn_4$
Tl		Mg_5Tl_2 Mg_2Tl $MgTl$	Ca_3Tl Ca_5Tl_2 Ca_3Tl_2 $CaTl$ Ca_3Tl_4 $CaTl_3$	Sr_3Tl Sr_5Tl_3 $SrTl$ Sr_2Tl_3 $SrTl_2$ $SrTl_3$	$Ba_{13}Tl$ Ba_2Tl $BaTl$ $BaTl_2$ $BaTl_3$ $BaTl_4$

B \ A	Zn	Cd	Hg
B		CdB_6?	
Al			
Ga			
In		Cd_3In	Hg_4In $HgIn$ $HgIn_2$
Tl			Hg_5Tl_2 Hg_7Tl_3 $HgTl$

The stoichiometry for homodesmic or simple binaries is predicted to be A_5B_2. Thus for homodesmic binaries: $b/a = e_A/v_B = 2/5$, for heterodesmic polyanionic binaries: $b/a > 2/5$, and for heterodesmic polycationic binaries: $b/a < 2/5$. As with the previous tables for the Group *2/6*-Group *4/4* binaries, there is a notable lack of compounds formed between the Group *3/5* elements and those elements in Group *2/6* having *FEN* values greater or equal to 0.28 (i.e. Be, Zn, Cd and Hg). Only a few of the observed compounds are simple binaries (Mg_5Ga_2, Mg_5In_2, Mg_5Tl_2, Ca_5Tl_2) and only a few correspond to polycationic binaries ((Ca_3)Tl, (Ba_{10})Ga, (Ba_{13})In, etc.), with the vast majority falling into the category of polyanionic binaries.

Though resulting bond parallelogram lies completely within the predominantly metallic region, the resulting compounds are once again best characterized as semiconductors rather than as pure metals. The ionic character of the bonds varies between 28% for the Ba-B bond and 3% for the Hg-In bond, whereas their metallic character varies between 62% for the boride bonds and 72% for the indide bonds. Note again the highly compressed form of the parallelogram reflecting the small range of *FEN* values for the

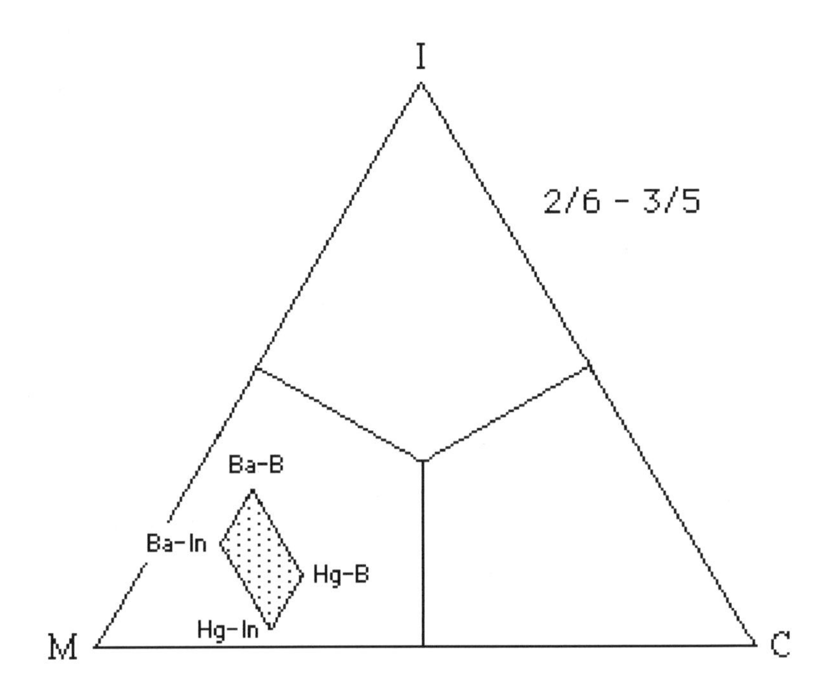

Group *3/5* elements due to the operation of both the transition-block and inner-transition block insertions.

6.7 Group *2/6* - Group *2/6* Binaries

The net stoichiometries of most of the known binary compounds formed among the Group *2/6* elements are summarized in the three tables at the bottom of this page and the top of the following page (7).

The stoichiometry for homodesmic or simple binaries is predicted to be A_3B. Thus for homodesmic binaries: $b/a = e_A/v_B = 2/6 = 1/3$, for heterodesmic polyanionic binaries: $b/a > 1/3$, and for heterodesmic polycationic binaries: $b/a < 1/3$. Compound formation is rare among the members of the Ca branch and among the members of the Zn branch, but is common for all other possible combinations. Only a few simple binaries (e.g., Mg_3Hg, Mg_3Cd, Ca_3Hg, Ca_3Zn, Ca_3Cd, Zn_3Hg) are known, with most of the remaining compounds falling into the category of polyanionic binaries. All of these compounds are generally considered to be metallic in character.

B \ A	Be	Mg	Ca	Sr	Ba
Be		$MgBe_{13}$	$CaBe_{13}$	$SrBe_{13}$	$BaBe_{13}$
Mg			$CaMg_2$	$SrMg_2$ Sr_6Mg_{23} Sr_2Mg_{17}	$BaMg_2$ Ba_6Mg_{23} Ba_2Mg_{17}
Ca					
Sr					
Ba					

B \ A	Be	Mg	Ca	Sr	Ba
Hg	$BeHg_2$	Mg_3Hg Mg_5Hg_2 Mg_2Hg Mg_3Hg_5 $MgHg$ $MgHg_2$	Ca_3Hg Ca_2Hg Ca_5Hg_3 Ca_3Hg_2 $CaHg$ $CaHg_2$ $CaHg_3$	Sr_3Hg Sr_2Hg Sr_3Hg_2 $SrHg$ $SrHg_2$ $SrHg_3$ Sr_2Hg_9 $SrHg_{11}$ $SrHg_{13}$	Ba_2Hg $BaHg$ $BaHg_2$ $BaHg_4$ Ba_2Hg_9 $BaHg_6$ $BaHg_{11}$ $BaHg_{13}$
Zn		Mg_7Zn_3 $MgZn$ Mg_2Zn_3 $MgZn_2$ Mg_2Zn_{11}	Ca_3Zn Ca_5Zn_3 $CaZn$ $CaZn_2$ $CaZn_3$ $CaZn_5$ $CaZn_{11}$ $CaZn_{13}$	$SrZn$ $SrZn_2$ $SrZn_5$ $SrZn_{13}$	Ba_2Zn $BaZn$ $BaZn_2$ $BaZn_5$ $BaZn_{13}$
Cd		Mg_3Cd $MgCd$ $MgCd_3$	Ca_3Cd_2 $CaCd$ $CaCd_2$ Ca_3Cd_{17} $CaCd_6$	Sr_2Cd $SrCd$ $SrCd_2$ Sr_2Cd_9 $SrCd_6$ $SrCd_{11}$	Ba_2Cd $BaCd$ $BaCd_2$ Ba_7Cd_{31} $BaCd_{11}$

B \ A	Hg	Zn	Cd
Hg		Zn_3Hg $ZnHg_3$	
Be			
Zn			
Cd			

Once again the typical bond parallelogram has been truncated to a triangle due to its intersection with the bottom edge of the bond-type triangle, thereby automatically eliminating redundant but inverted bond combinations. The ionic character of the bonds ranges from a maximum of 15% for Ba-Hg to a minimum of 0% for all of the self A-A bonds lying along the bottom edge of the triangle, whereas the metallic character varies from a maximum of 90% for the Ba bonds to a minimum of 75% for the Hg bonds. Likewise the covalent character of the simple substances lying along the base varies from 10% for the Ba-Ba bonds to 25% for the Hg-Hg bonds.

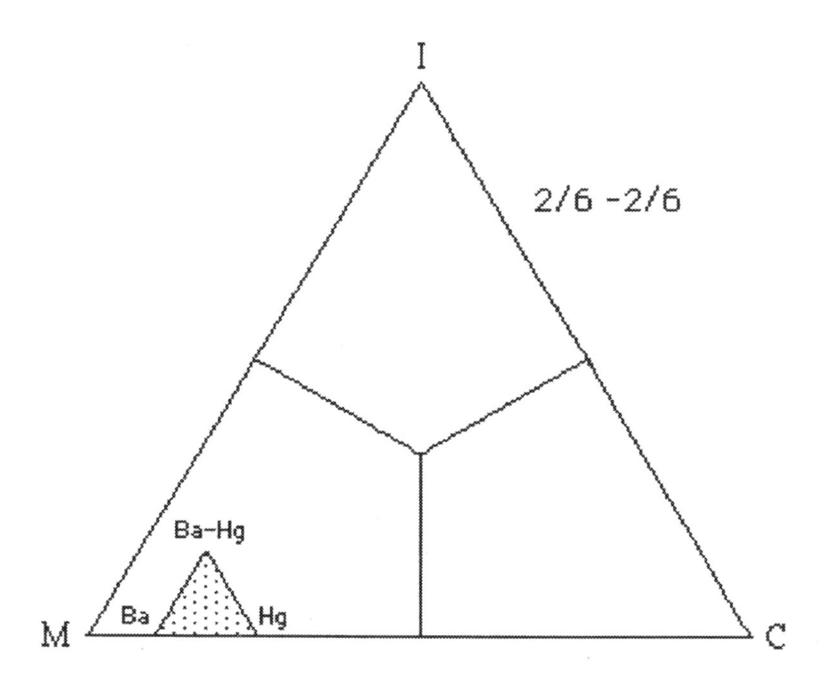

6.8 References and Notes

1. H. Schäfer, B. Eisenmann, W. Müller, "Zintl Phases: Transitions Between Metallic and Ionic Bonding," *Angew. Chem. Int. Ed. Engl.*, **1973**, *12*, 694-712.

2. W. B. Jensen, "The Place of Zinc, Cadmium, and Mercury in the Periodic Table," *J. Chem. Educ.*, **2003**, *80*, 952-961.

3. W. Klemm, E. Busmann, "Volumeninkremente und Radien einiger einfach negativ geladener Ionen," *Z. Anorg. Allg. Chem.*, **1963**, *319*, 297-311.

4. H. A. Bent, "Isoelectronic Systems," *J. Chem. Educ.*, **1966**, *43*, 170-186.

5. .N. Wiberg, *Inorganic Chemistry*, Academic Press: New York, NY, 2001, Chapter XVII.

6. F. Basolo, "Stabilization of Metal Complexes by Large Counterions," *Coord. Chem. Rev.*, **1968**, *3*, 213-223.

7. B. Predel, Ed., *Phase Equilibria, Crystallographic and Thermodynamic Data of Binary Alloys,* Landolt-Börnstein, Vols. 5a - 5i, Springer: Berlin, 1991-1998.

8. N. N. Greenwood, A. Earnshaw, *Chemistry of the Elements*, Pergamon: Oxford, 1994, p. 1403.

VII

Bonding in the Binary Compounds of Groups *3/5* and *4/4*

7.1 Preliminary Observations

The A-B bonds formed by Groups *1/7* and *2/6* shared a common progression from the predominantly ionic to the predominantly metallic. In the case of the A-B bonds formed by Groups *3/5* and *4/4* one finds that this gradual transformation now occurs as a three-stage, rather than a two-stage, progression, which passes from predominantly ionic to predominantly metallic via a partial intersection with the predominantly covalent region. Because of the irregular trends in *FEN* values for the Group *3/5* elements, due to the operation of both the transition-block and inner-transition block insertions, the members of this group have been placed in the binary compound tables in order of their decreasing *FEN* values (B, Ga, Tl, Al, In) rather than their increasing atomic numbers (B, Al, Ga, In, Tl), as this more closely correlates with the order in which their A-B bonds appear in the corresponding bond triangles.

Since our bond survey for a given group of the main-block elements involves the only those compounds formed either amongst the members of the group itself or with the members of more electronegative groups that are found to its right in the periodic table, the number of nonredundant A-B combinations we need to consider decreases as we move from left to right across the main block. Hence, for what remains, we are able to survey the combinations of more than one group in a given lecture.

7.2 Group *3/5* - Group *7/1* Binaries

The net stoichiometries of most of the known binary compounds formed between the Group *3/5* elements and the halogens or Group *7/1* elements are summarized in the following table (1, 2):

B \ A	B	Ga	Tl	Al	In
F	BF B_2F_4 BF_3	GaF_3	TlF TlF_3	AlF_3	InF InF_3
Cl	$(BCl)_n$ (n = 4, 8-12) B_2Cl_4 BCl_3	Ga_2Cl_3 $GaCl_2$ Ga_3Cl_7 $GaCl_3$	TlCl Tl_2Cl_3 $TlCl_2$ $TlCl_3$	$AlCl_3$	InCl In_3Cl_4 In_2Cl_3 In_5Cl_9 $InCl_3$
Br	$(BBr)_n$ (n = 7-10) B_2Br_4 BBr_3	$GaBr_2$ $GaBr_3$	TlBr Tl_2Br_3 $TlBr_2$ $TlBr_3$	$AlBr_3$	InBr In_2Br_3 $InBr_2$ In_4Br_7 $InBr_3$
I	$(BI)_n$ (n = 8-9) B_2I_4 BI_3	GaI_2 GaI_3	TlI TlI_3	AlI_3	InI InI_2 InI_3

The stoichiometry for homodesmic or simple binaries is predicted to be AB_3. Thus for homodesmic binaries: $b/a = e_A/v_B = 3/1$, for heterodesmic polyanionic binaries: $b/a > 3/1$, and for heterodesmic polycationic binaries: $b/a < 3/1$. Examples of simple binaries are known for all of the resulting A-B bonds, with the rest of the compounds falling into the category of polycationic. Here for the first time we also encounter two new phenomena:

1. All of the binary compounds reported in the previous two lectures corresponded to infinitely extended polymers or crystals. Here for the first time we begin to encounter discrete molecular species. Thus the relative stoichiometric formula $AlBr_3$ corresponds to the absolute molecular formula Al_2Br_6, BF_3 is both the relative stoichiometric formula and the absolute molecular formula of boron trifluoride, etc. Hence, from this point on, either net stoichiometric or absolute molecular formulas may appear in the

compound tables, depending on circumstances.

2. For the first time we begin encounter polycationic compounds which contain cationic lone pairs rather than A-A bonds. Thus, unlike BCl, which contains B-B bonds, TlCl contains no Tl-Tl bonds, but rather Tl^{+} ions and, from this point on, the term polycationic must be taken to imply either of these two bonding alternatives.

Though the resulting bond parallelogram places most of these bonds in the predominantly ionic region, this must not be taken to imply that the resulting compounds automatically have the properties of typical ionic salts. Rather in this case they have properties best interpreted as belonging to species with highly polar covalent bonds in which the ionic character ranges from a maximum of 72% for the In-F bond to a minimum of 24% for the B-I bond, and the metallic character ranges from 0% by definition for the fluoride bonds to a maximum of 38% for the iodide bonds. Because of the narrow range in *FEN* values for Group 3/5, due to the effects of both the transition block and inner-transition block insertions, covalency values vary

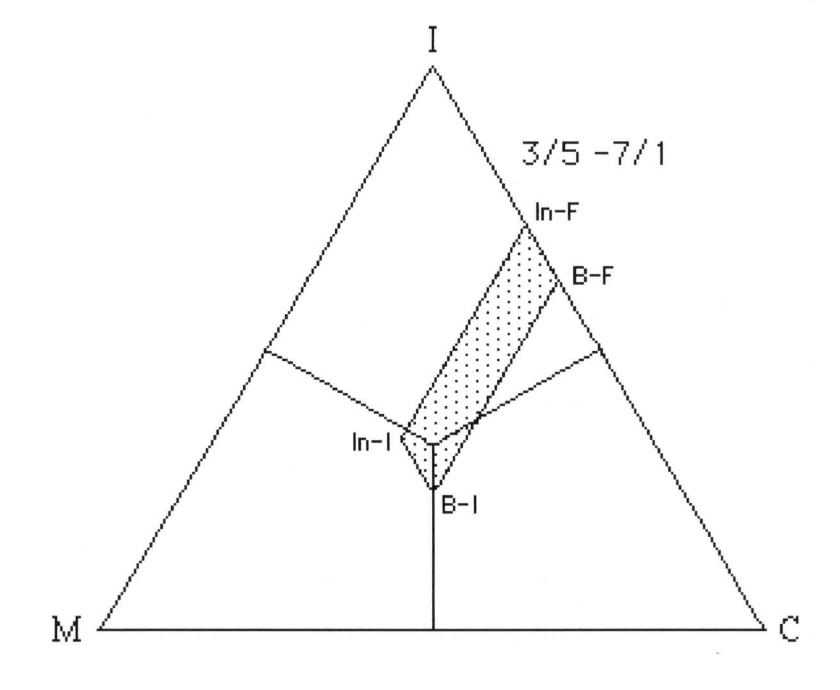

by only 10% on passing from the In to the B bonds and the bond parallelogram is correspondingly quite narrow.

7.3 Group *3/5* - Group *6/2* Binaries

The net stoichiometries of most of the known binary compounds formed between the Group *3/5* elements and the chalcogens or Group *6/2* elements are summarized in the following table (1, 2).

B \ A	B	Ga	Tl	Al	In
O	B_7O B_6O B_2O B_4O_3 BO B_4O_5 B_2O_3	Ga_2O Ga_2O_3	Tl_2O Tl_3O_4 Tl_2O_3	Al_2O AlO Al_2O_3	In_2O InO In_2O_3
S	$B_{12}S$ B_4S BS B_2S_3 BS_2	Ga_2S GaS Ga_4S_5 Ga_2S_3	Tl_2S Tl_4S_3 TlS Tl_2S_3 Tl_2S_5	AlS Al_2S_3	InS In_6S_7 In_2S_3
Se	B_2Se_3	Ga_2Se $GaSe$ Ga_2Se_3	Tl_2Se $TlSe$ Tl_2Se_3	Al_2Se_3	In_4Se_3 $InSe$ In_6Se_7 In_2Se_3
Te		$GaTe$ Ga_3Te_4 Ga_2Te_3 Ga_2Te_5	Tl_5Te_3 $TlTe$ Tl_2Te_3	Al_2Te_3	In_4Te_3 $InTe$ In_3Te_4 In_2Te_3 In_2Te_5

The stoichiometry for homodesmic or simple binaries is predicted to be A_2B_3. Thus for homodesmic binaries: $b/a = e_A/v_B = 3/2$, for heterodesmic polyanionic binaries: $b/a > 3/2$, and for heterodesmic polycationic binaries: $b/a < 3/2$. As may be seen, save for the B-Te systems, for which no binary compounds are apparently known, all of the bonds may be illustrated using simple binaries.

As shown at the top of the following page, the parallelogram for these bonds, which overlaps with that for the halide bonds, lies partially within the predominantly ionic region, partially within the predominantly metallic region, and partially within the predominantly covalent region, with the ionic character ranging from a maximum of 57% for the In-O bond to a minimum of 16%

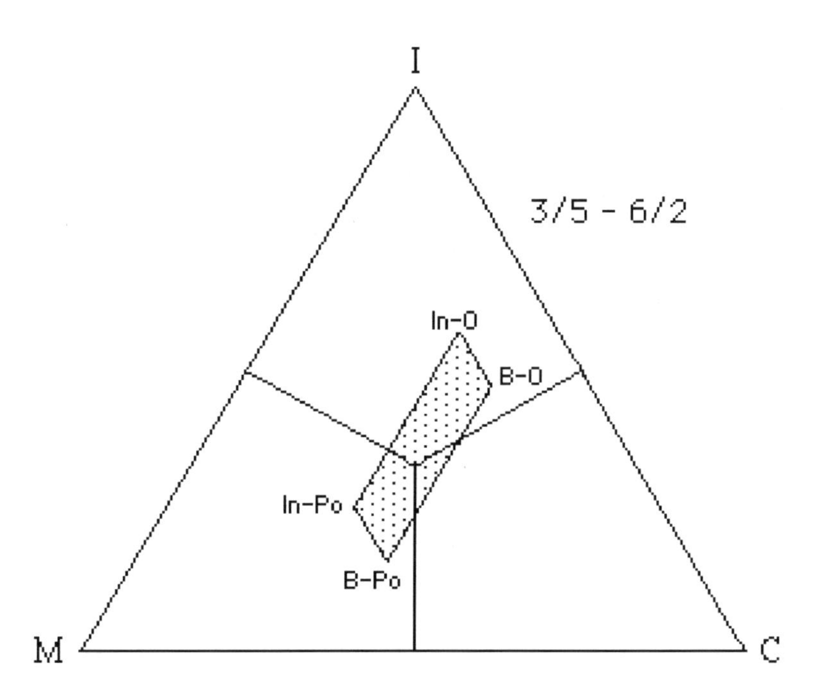

for the B-Po bond. As before, the covalency variation remains at 10%, whereas the metallicity varies from a minimum of 15% for the oxide bonds to a maximum of 46% for the polonide bonds.

7.4 Group *3/5* - Group *5/3* Binaries

The net stoichiometries of most of the known binary compounds formed between the Group *3/5* elements and the pnictogens or Group *5/3* elements are summarized the table at the top of the following page (1, 2).

The stoichiometry for homodesmic or simple binaries is predicted to be AB. Thus for homodesmic binaries: $b/a = e_A/v_B = 3/3 = 1/1$, for heterodesmic polyanionic binaries: $b/a > 1/1$, and for heterodesmic polycationic binaries: $b/a < 1/1$. Save for the B-Bi Al-Bi, Ga-Bi systems, for which no binary compounds are apparently known, all of the bonds may be illustrated using simple binaries. The only polyanionic species are $Tl(P_5)$ and the azides of boron, $BN_9 = B(N_3)_3$, as well as both $TlN_3 = Tl(N_3)$ containing Tl^+, and the mixed valence $TlN_6 = (Tl^+Tl^{3+})(N_3)_4$.

B \ A	B	Ga	Tl	Al	In
N	$B_{50}N$ $B_{24}N$ B_3N BN BN_9	GaN	TlN TlN_3 TlN_6	AlN	InN
P	B_6P $B_{13}C_2$ BP	GaP	Tl_3P TlP TlP_5	AlP	InP
As	B_6As BAs	GaAs	Tl_3As	AlAs	InAs
Sb	BSb	GaSb	Tl_3Sb TlSb	AlSb	InSb
Bi			Tl_3Bi Tl_2Bi_3 $TlBi_2$		In_2Bi In_5Bi_3 InBi

The resulting bond parallelogram, which overlaps with that for the chalcide bonds, places the In-N bond in the predominantly ionic region, the B-N bond in the predominantly covalent region, and the rest of the bonds in the predominantly metallic region, with the ionicity ranging from a maximum

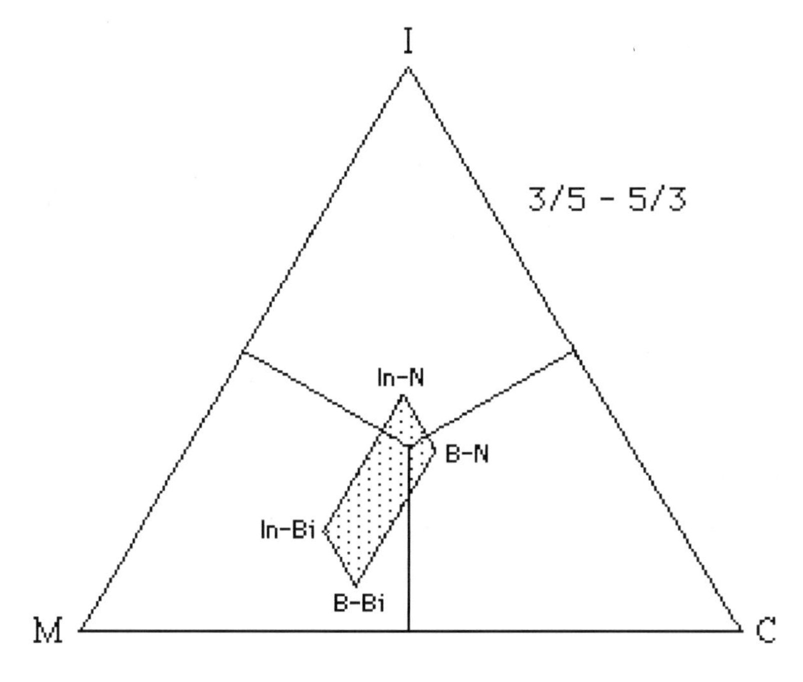

of 42% for the In-N bond to a minimum of 7% for the hypothetical B-Bi bond, and the metallicity from a maximum of 54% for the bismuthide bonds to a minimum of 30% for the nitride bonds. The placement of the B-N bond in the predominantly covalent region is in keeping with the well-known iso-electronic analogy between the chemistry of this bond and that of the C-C bond (3), and of course, many of the A-B bonds falling in the predominantly metallic region are associated with simple binaries, such as AlP, InSb, etc, which have long been acknowledged as quintessential examples of simple semiconductors (4).

7.5 Group *3/5* - Group *4/4* Binaries

The net stoichiometries of most of the known binary compounds formed between the Group *3/5* elements and the carbon or Group *4/4* elements are summarized in the following table (1, 2):

B \ A	B	Ga	Tl	Al	In
C	$B_{50}C$ $B_{50}C_2$ $B_{13}C_2$			$\underline{Al_4C_3}$	
Si	$B_{14}Si$ B_6Si B_3Si				
Ge					
Sn			TlSn		
Pb					

The stoichiometry for homodesmic or simple binaries is predicted to be A_4B_3. Thus for homodesmic binaries: $b/a = e_A/v_B = 3/4$, for heterodesmic polyanionic binaries: $b/a > 3/4$ and for heterodesmic polycationic binaries: $b/a < 3/4$. The most striking feature of this table is the large number of systems for which no binary compounds have been reported in either the inorganic textbook literature or the phase literature, many of these systems being either immiscible or forming solid solutions of variable composition instead. The only simple binary is apparently aluminum carbide.

The resulting, apparently largely hypothetical, bond parallelogram now lies completely within the predominantly metallic region, and the missing binary compounds, should they ever be prepared, would probably best be described as semiconductors rather than as true metals. The ionic character of the bonds would vary between 26% for the In-C bond (5) and 0% for the apparently hypothetical B-Pb bond, whereas their metallic character would vary between 46% for the carbide bonds and 62% for the plumbide bonds.

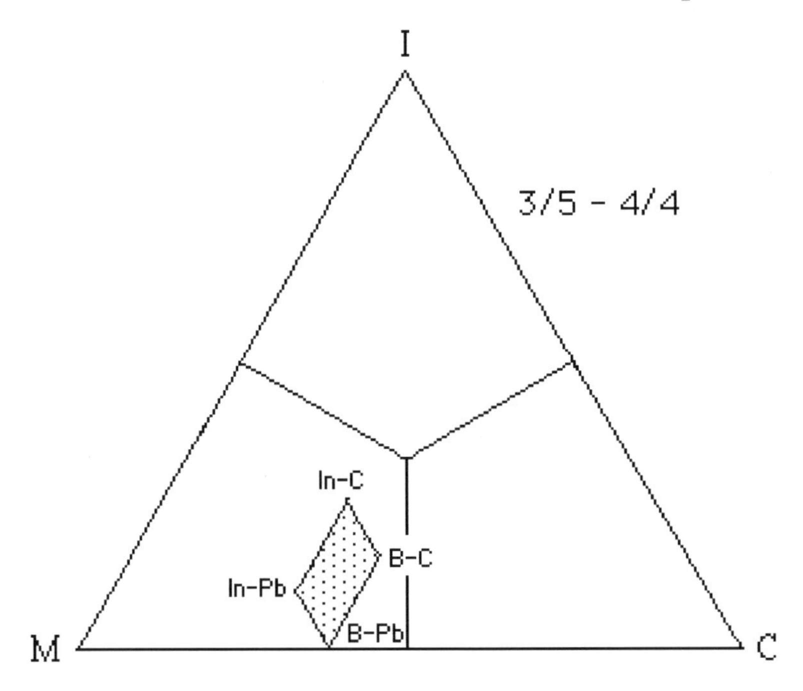

7.6 Group 3/5 - Group 3/5 Binaries

The net stoichiometries of most of the known binary compounds formed among the Group 3/5 elements are summarized in the following table: (1, 2):

B \ A	B	Ga	Tl	Al	In
B		GaB_{12} ?		AlB_2 AlB_{12}	
Ga					
Tl					
Al					
In					

The stoichiometry for homodesmic or simple binaries is predicted to be A_5B_3. Thus for homodesmic binaries: $b/a = e_A/v_B = 3/5$, for heterodesmic polyanionic binaries: $b/a > 3/5$, and for heterodesmic polycationic binaries: $b/a < 3/5$. As with the previous table, both the inorganic textbook literature and the phase literature report an almost total absence of binary compounds for these systems, with either immiscibility or solid solution formation being the norm. No simple binaries are known.

As usual, the bond parallelogram has been truncated to a triangle due to its intersection with the bottom edge of the bond-type triangle, thereby automatically eliminating redundant but inverted bond combinations. The ionic character of the bonds ranges from a maximum of 10% for the In-B bond to a minimum of 0% for all of the self A-A bonds lying along the bottom edge of the triangle, whereas the metallic character varies from a maximum of 72% for the indide bonds to a minimum of 62% for the boride bonds. Likewise the covalent character of the simple substances lying along the base varies from 38% for the B-B bonds to 28% for the In-In bonds.

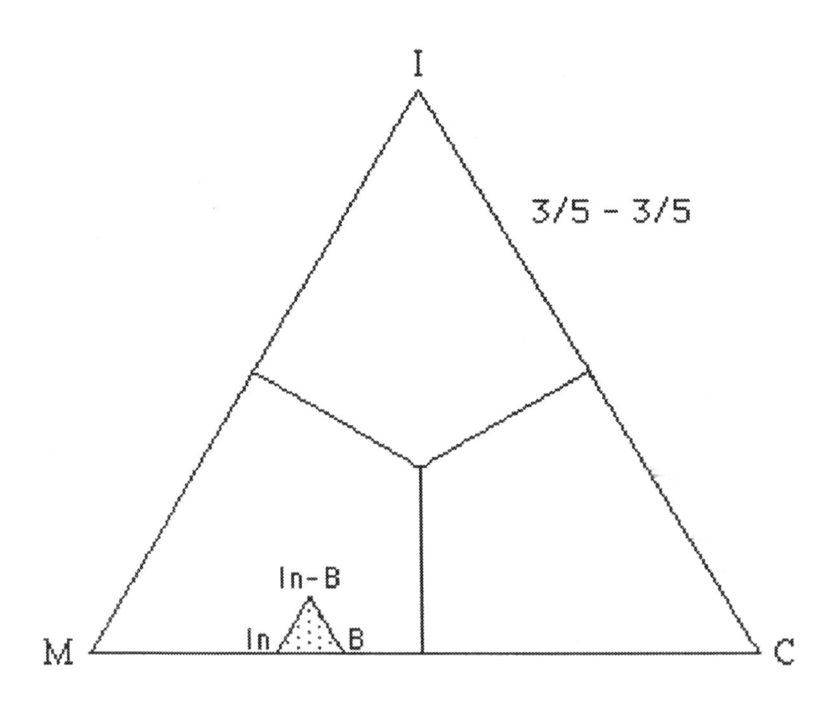

7.7 Group *4/4* - Group *7/1* Binaries

The net stoichiometries of most of the known binary compounds formed between the Group *4/4* elements and the halogens or Group *7/1* elements are summarized in the following table (1, 2):

B \ A	C	Si	Ge	Sn	Pb
F	C_aF_b (b/a < 4) $\underline{CF_4}$	Si_aF_b (b/a < 4) $\underline{SiF_4}$	GeF_2 $\underline{GeF_4}$	SnF_2 $\underline{SnF_4}$	PbF_2 $\underline{PbF_4}$
Cl	C_aCl_b (b/a < 4) $\underline{CCl_4}$	Si_aCl_b (b/a < 4) $\underline{SiCl_4}$	$GeCl_2$ Ge_aCl_{2a+2} $\underline{GeCl_4}$	$SnCl_2$ $\underline{SnCl_4}$	$PbCl_2$ $\underline{PbCl_4}$
Br	C_aBr_b (b/a < 4) $\underline{CBr_4}$	Si_aBr_b (b/a < 4) $\underline{SiBr_4}$	$GeBr_2$ $\underline{GeBr_4}$	$SnBr_2$ $\underline{SnBr_4}$	$PbBr_2$
I	C_aI_b (b/a < 4) $\underline{CI_4}$	Si_aI_b (b/a < 4) $\underline{SiI_4}$	GeI_2 $\underline{GeI_4}$	SnI_2 $\underline{SnI_4}$	PbI_2

The stoichiometry for homodesmic or simple binaries is predicted to be AB_4. Thus for homodesmic binaries: $b/a = e_A/v_B = 4/1$, for heterodesmic polyanionic binaries: $b/a > 4/1$, and for heterodesmic polycationic binaries: $b/a < 4/1$. Since in principle it possible to prepare a fully halogenated analog for any of the known hydrocarbons, the number and possible stoichiometries for the polycationic carbon halides is literally legion and consequently they have been represented using generalized formulas in the table, the same being true to a lesser extent of the silanes and the corresponding silicon halides. With the exception of the heavier lead halides, all of the bonds can be illustrated using simple binaries.

The resulting bond parallelogram lies mostly within the predominantly covalent region, with the ionicity varying from a maximum of 62% for the Pb-F bond to a minimum of 8% for the C-I bond and the metallicity from a maximum of 38% for the iodide bonds to a minimum of 0% by definition for the fluoride bonds. The covalency varies from a maximum of 54% for the carbon bonds to a minimum of 38% for the lead bonds.

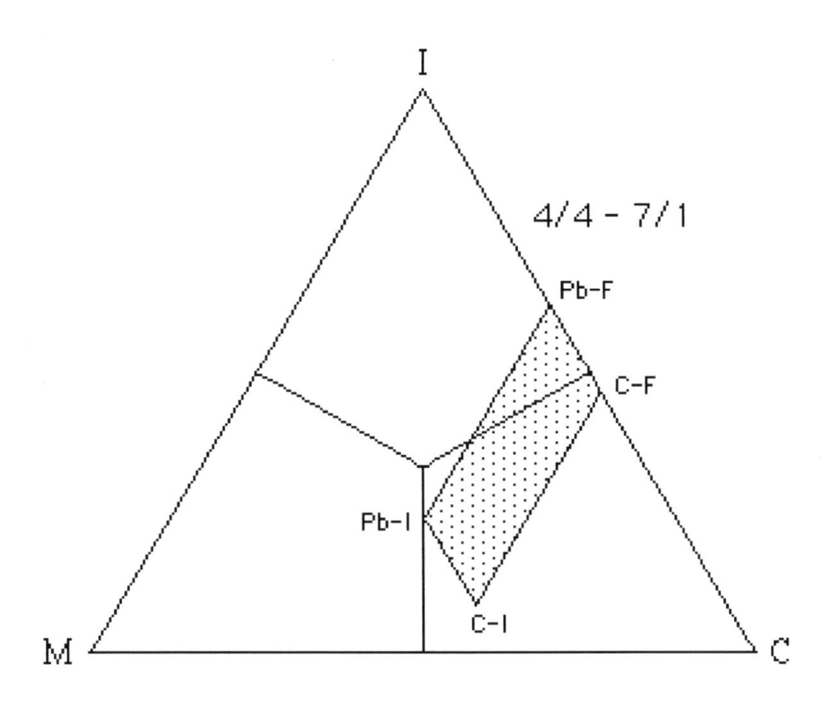

7.7 Group *4/4* - Group *6/2* Binaries

The net stoichiometries of most of the known binary compounds formed between the Group *4/4* and the chalcogens or Group *6/2* elements are summarized in the following table (1, 2):

B \ A	C	Si	Ge	Sn	Pb
O	C_3O_2 $C_{12}O_9$ CO $\underline{CO_2}$	SiO $\underline{SiO_2}$	GeO $\underline{GeO_2}$	SnO Sn_3O_4 $\underline{SnO_2}$	PbO Pb_3O_4 $\underline{PbO_2}$
S	C_3S_2 CS C_9S_9 C_4S_6 $\underline{CS_2}$	SiS $\underline{SiS_2}$	GeS $\underline{GeS_2}$	SnS Sn_2S_3 $\underline{SnS_2}$	PbS
Se	$\underline{CSe_2}$	$\underline{SiSe_2}$	GeSe $\underline{GeSe_2}$	SnSe Sn_2Se_3 $SnSe_3$	PbSe
Te		Si_2Te_3	GeTe	SnTe	PbTe

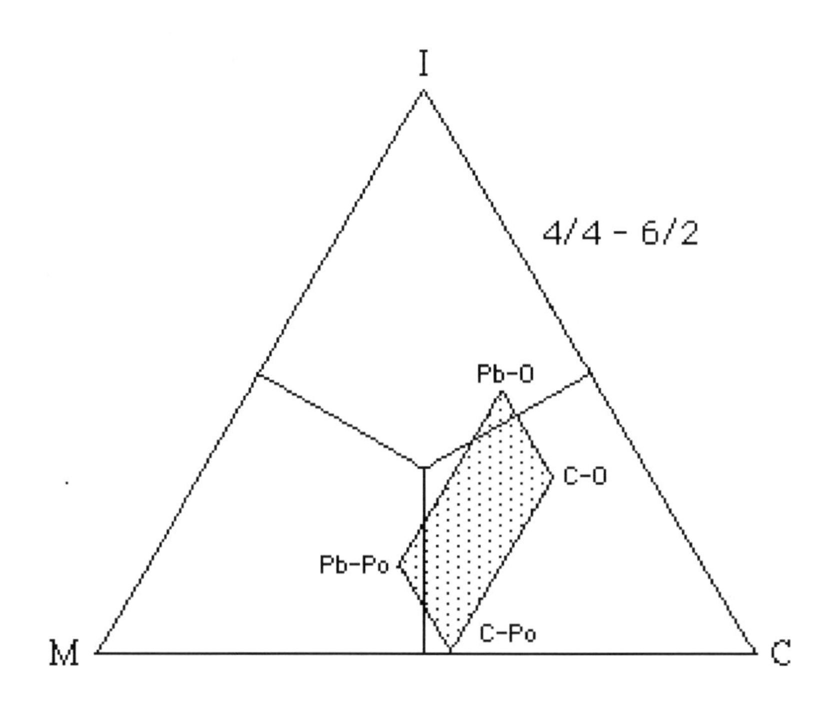

The stoichiometry for homodesmic or simple binaries is predicted to be AB_2. Thus for homodesmic binaries: $b/a = e_A/v_B = 4/2 = 2/1$, for heterodesmic polyanionic binaries: $b/a > 2/1$, and for heterodesmic polycationic binaries: $b/a < 2/1$. A fair number of simple binaries are known, though there appears to be no available data on the C-Te system. Most of the remaining binaries fall into the polycationic category.

The resulting bond parallelogram (see bottom of previous page) lies mostly within the predominantly covalent region, with the ionicity varying from a maximum of 47% for the Pb-O bond to a minimum of 0% for the C-Po Bond and the metallicity from a maximum of 48% for the polonide bonds to a minimum of 15% for the oxide bonds.

7.8 Group *4/4* - Group *5/3* Binaries

The net stoichiometries of most of the known binary compounds formed between the Group *4/4* and the pnictogens or Group *5/3* elements are summarized in the following table (2):

B \ A	C	Si	Ge	Sn	Pb
N	CN	Si_3N SiN Si_3N_4 SiN_{12}	Ge_3N_4	Sn_3N_4	PbN_6
P		SiP SiP_2?	GeP GeP_3 GeP_5	SnP Sn_4P_3 Sn_3P_4 SnP_3	Pb_6P
As		$SiAs$ $SiAs_2$	$GeAs$ $GeAs_2$	$SnAs$ Sn_4As_3	
Sb				Sn_3Sb_2 $SnSb$	
Bi					Pb_3Bi_5 Pb_3Bi_7 $PbBi_9$

The stoichiometry for homodesmic or simple binaries is predicted to be A_3B_4. Thus for homodesmic binaries: $b/a = e_A/v_B = 4/3$, for heterodesmic polyanionic binaries: $b/a > 4/3$, and for heterodesmic polycationic binaries: $b/a < 4/3$. Only four examples of simple binaries are known and for many of the systems there are apparently no reported binary compounds at all.

The resulting bond parallelogram lies mostly within the predominantly covalent region, with the ionicity varying from a maximum of 32% for the Pb-N bond to a minimum of 0% for the A-A bonds and the metallicity from a maximum of 56% for the bismuthide bonds to a minimum of 30% for the nitride bonds. Since the *FEN* value for C is greater than those for P, As, Sb, and Bi, its binary compounds with these elements are formally carbides rather than phosphides, arsenides, etc. as indicated by both the partial truncation of the bond parallelogram, due to its intersection with the bottom of the triangle, and the shaded boxes in the above table. As a consequence, these combinations will appear instead in the survey of the binary compounds of Group *5/3* found in the next lecture.

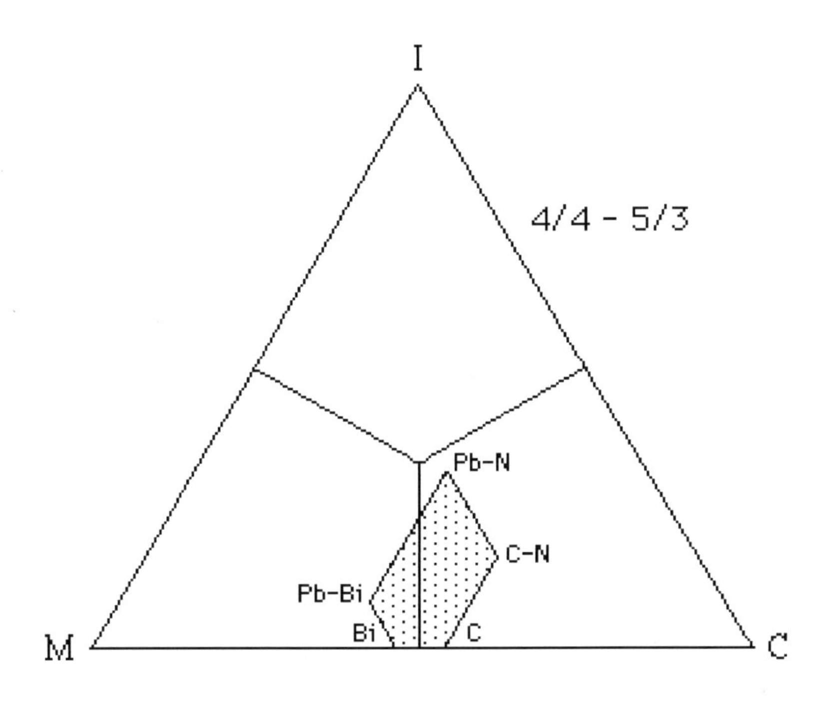

7.9 Group *4/4* - Group *4/4* Binaries

The net stoichiometries of most of the known binary compounds formed among the Group *4/4* elements are summarized in the following table (1, 2):

B \ A	C	Si	Ge	Sn	Pb
C		SiC			PbC_2
Si					
Ge					
Sn					
Pb					

The stoichiometry for homodesmic or simple binaries is predicted to be AB. Thus for homodesmic binaries: $b/a = e_A/v_B = 4/4 = 1/1$, for heterodesmic polyanionic binaries: $b/a > 1/1$, and for heterodesmic polycationic binaries: $b/a < 1/1$. As with the earlier table for the possible intragroup combinations for Group *3/5*, one is struck by the lack of binary compounds reported in either the inorganic textbook literature or in the phase-diagram literature, with complete immiscibility or solid solution formation being the norm instead. Only one simple binary (SiC) is listed, albeit one which is both well-known and of great industrial importance.

As usual, the bond parallelogram has been truncated to a triangle due to its intersection with the bottom edge of the bond-type triangle, thereby automatically eliminating redundant but inverted bond combinations. The ionic character of the bonds ranges from a maximum of 16% for the Pb-C bond to a minimum of 0% for all of the self A-A bonds lying along the bottom edge of the triangle, whereas the metallic character varies from a maximum of 62% for the plumbide bonds to a minimum of 46% for the carbide bonds. Likewise the covalent character of the simple substances lying along the base varies from 54% for the C-C bonds to 38% for the Pb-Pb bonds.

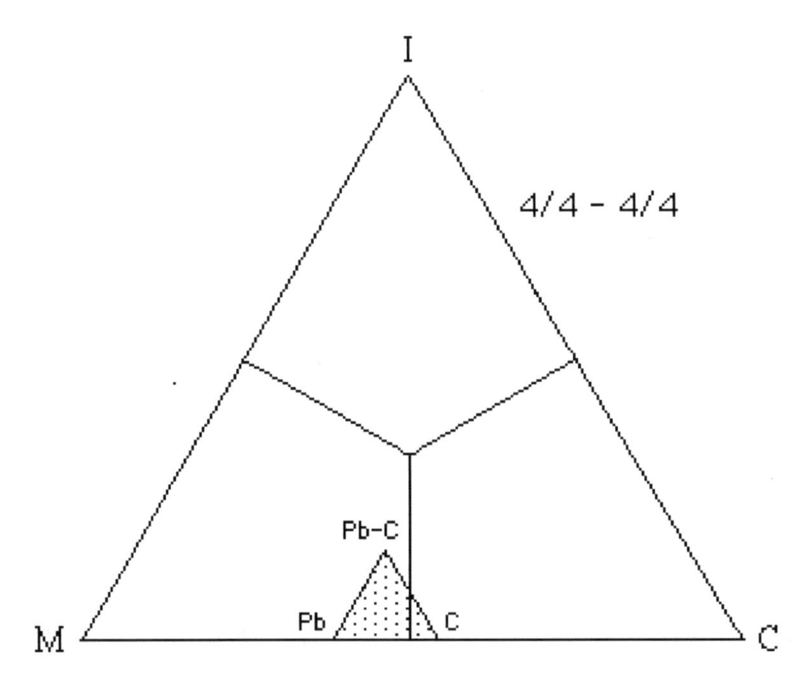

7.10 References and Notes

1. N. Wiberg, *Inorganic Chemistry*, Academic Press: New York, NY, 2001, Chapters XV-XVI.

2. B. Predel, Ed., *Phase Equilibria, Crystallographic and Thermodynamic Data of Binary Alloys,* Landolt-Börnstein, Vols. 5a-5i Springer: Berlin, 1991-1998.

3. K. Niedenzu, J. W. Dawson, *Boron-Nitrogen Compounds*, Academic Press: New York, NY, 1965.

4 N. A. Goryunova, *The Chemistry of Diamond-Like Semiconductors*, MIT Press: Cambridge, MA, 1963.

5. Many examples of In-C bonds can, of course, be found among its various organometallic compounds. See C. Elschenbroich, A. Salzer, *Organometallics: A Concise Introduction*, 2nd ed., VCH: Weineheim, 1992, pp. 86-92.

VIII
Bonding in the Binary Compounds of Groups *5/3*, *6/2*, *7/1* and *8/0*

8.1 Preliminary Observations

The A-B bonds formed by Groups *1/7* and *2/6* shared a common progression from the predominantly ionic to predominantly metallic. In the case of the A-B bonds formed by Groups *3/5* and *4/4* this progression occurred in three rather than two stages, passing from predominantly ionic to the predominantly metallic via a partial intersection with the predominantly covalent region. Finally, in the case of the A-B bonds formed by Groups *5/3* through 8/0, discussed in this lecture, we find that virtually all of the bonds fall into the predominantly covalent region, with only a small residual overlap with the predominantly ionic and metallic regions in the case of Group *5/3*. This means that an increasing number of the resulting binary compounds have discrete molecular, rather than infinitely polymerized, structures and that, consequently, an increasing number of the formulas in the tables are absolute rather than relative in nature. In addition, for the first time we also encounter two additional phenomena:

1. None of the resulting bond parallelograms are complete. In other words, all of them exhibit some degree of truncation due to their intersection with the base of the bond triangle.

2. For the first time we must consider not only the A-B bonds formed between a given group and the groups lying to its right in the periodic table,

but also those involving the first member of the group to its immediate left. Both of these phenomena are due to the fact that the abnormally high *FEN* values of C, N, O and F overlap with those of the heavier elements in the groups to their immediate right, whereas there was no intergroup overlap between the *FEN* values of the elements in Groups *1/7* through *4/4* discussed in the previous three lectures. This overlap also means that the corresponding truncated bond parallelograms also exhibit a substantial overlap with one another.

8.2 Group *5/3* - Group *7/1* Binaries

The net stoichiometries of most of the known binary compounds formed between the Group *5/3* elements and the halogens or Group *7/1* elements are summarized in the following table (1).

B \ A	N	P	As	Sb	Bi
F	N_3F N_2F_2 N_2F_4 NF_3	P_4F_6 P_2F_4 PF_3 $\underline{PF_5}$	AsF_3 $\underline{AsF_5}$	SbF_3 $\underline{SbF_5}$	BiF_3 $\underline{BiF_5}$
Cl	N_3Cl NCl_3	P_6Cl_6 P_2Cl_4 PCl_3 $\underline{PCl_5}$	$AsCl_3$ $\underline{AsCl_5}$	$SbCl_3$ $\underline{SbCl_5}$	$BiCl$ $BiCl_3$
Br		P_6Br_6 P_2Br_4 PBr_3 $\underline{PBr_5}$	$AsBr_3$	$SbBr_3$	$BiBr$ $BiBr_3$
I		P_2I_4 PI_3 $\underline{PI_5}$	As_2I_4 AsI_3	SbI_3	BiI BiI_3

The stoichiometry for homodesmic or simple binaries is predicted to be AB_5. Thus for homodesmic binaries: $b/a = e_A/v_B = 5/1$, for heterodesmic polyanionic binaries: $b/a > 5/1$, and for heterodesmic polycationic binaries $b/a < 5/1$. As may be seen, simple binaries are missing for N and for the heavier halides of As, Sb, and Bi. Polycationic binaries containing either A-A bonds (e.g. P_2Cl_4) or cationic lone pairs (e.g., PCl_3) are known. On the other hand, there are apparently no known examples of polyanionic binaries.

With the exception of the Bi-F bond, the resulting bond parallelogram places all of these bonds in the predominantly covalent region. Since the *FEN* value of N is greater than those of Br and I, intersection of the bond parallelogram with the base eliminates these bonds, which appear instead in the corresponding bond triangle for the Group *7/1*-Group *5/3* bonds as halogen nitrides. The ionic character ranges from a maximum of 54% for the Bi-F bond to a minimum of 0% for the A-A bonds along the base, and the metallic character ranges from 0% by definition for the fluoride bonds to a maximum of 38% for the iodide bonds, while the covalency ranges from a maximum of 70% for the nitrogen bonds to a minimum of 46% for the bismuth bond.

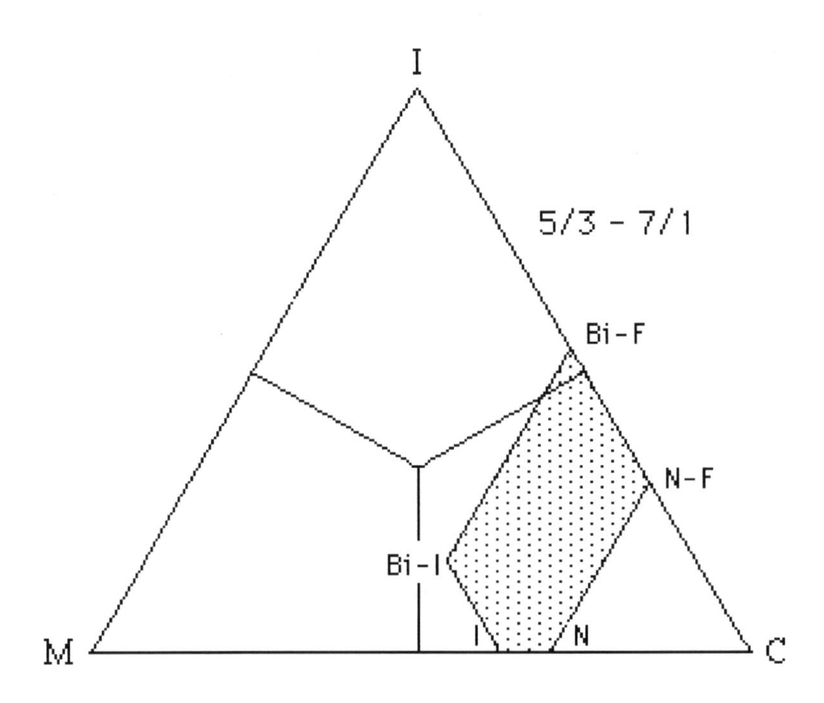

8.3 Group *5/3* - Group *6/2* Binaries

The net stoichiometries of most of the known binary compounds formed between the Group *5/3* elements and the chalcogens or Group *6/2* elements are summarized in the following table (1, 2):

B \ A	N	P	As	Sb	Bi
O	N_2O NO N_2O_3 NO_2 N_2O_5 NO_3	P_2O PO PO_2 P_4O_b (b = 6-10)	As_2O_3 As_2O_4 As_2O_5	Sb_2O_3 Sb_2O_4 Sb_2O_5	Bi_2O_3 Bi_2O_5
S		P_4S_b (b = 3-10)	As_4S_b (b = 3-6, 10)	Sb_2S_3 Sb_2S_5	Bi_2S_3
Se		P_4Se_3 P_4Se_4 P_2Se_5	AsSe As_2Se_3	Sb_2Se_3	Bi_3Se_2 BiSe Bi_2Se_3
Te			As_2Te_3 $AsTe_2$	Sb_2Te_3	$Bi_{14}Te_6$ Bi_2Te Bi_4Te_3 BiTe Bi_2Te_3

The stoichiometry for homodesmic or simple binaries is predicted to be A_2B_5. Thus for homodesmic binaries: $b/a = e_A/v_B = 5/2$, for heterodesmic polyanionic binaries: $b/a > 5/2$, and for heterodesmic polycationic binaries: $b/a < 5/2$. As may be seen, simple binaries are known for the oxides, a few sulfides, and for P_2Se_5, with the remaining compounds falling into the polycationic category.

The resulting bond parallelogram places all of these bonds in the predominantly covalent region. Since the *FEN* value of N is greater than those of S, Se, Te and Po, intersection of the bond parallelogram with the base eliminates these bonds, which appear instead in the corresponding bond triangle for the Group *6/2* - Group *5/3* bonds as chalcogen nitrides. The ionic character ranges from a maximum of 39% for the Bi-O bond to a minimum of 0% for the A-A bonds along the base, and the metallic character ranges from 15% for the oxide bonds to a maximum of 46% for the polonide bonds, while the covalency again ranges from a maximum of 70% for the nitrogen bonds to a minimum of 46% for the bismuth bonds.

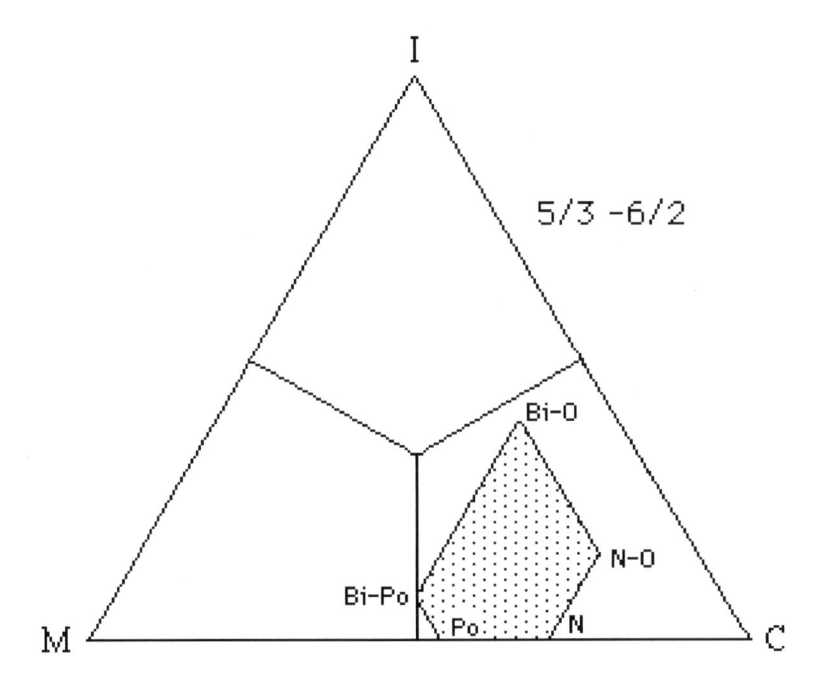

8.4 Group *5/3* - Group *5/3* Binaries

The net stoichiometries of most of the known binary compounds formed among the Group *5/3* elements are summarized in the table at the top of the following page (1, 2). The stoichiometry for homodesmic or simple binaries is predicted to be A_3B_5. Thus for homodesmic binaries: $b/a = e_A/v_B = 5/3$ for heterodesmic polyanionic binaries: $b/a > 5/3$, and for heterodesmic polycationic binaries: $b/a < \mathbf{5/3}$. Only the P-N system appears to be well characterized and provides examples of all three classes of compounds.

As per usual, intersection of the resulting bond parallelogram with the base eliminates redundant bond combinations. The ionic character ranges from a maximum of 24% for the Bi-N bond to a minimum of 0% for the A-A bonds along the base, and the metallic character ranges from a minimum of 30%

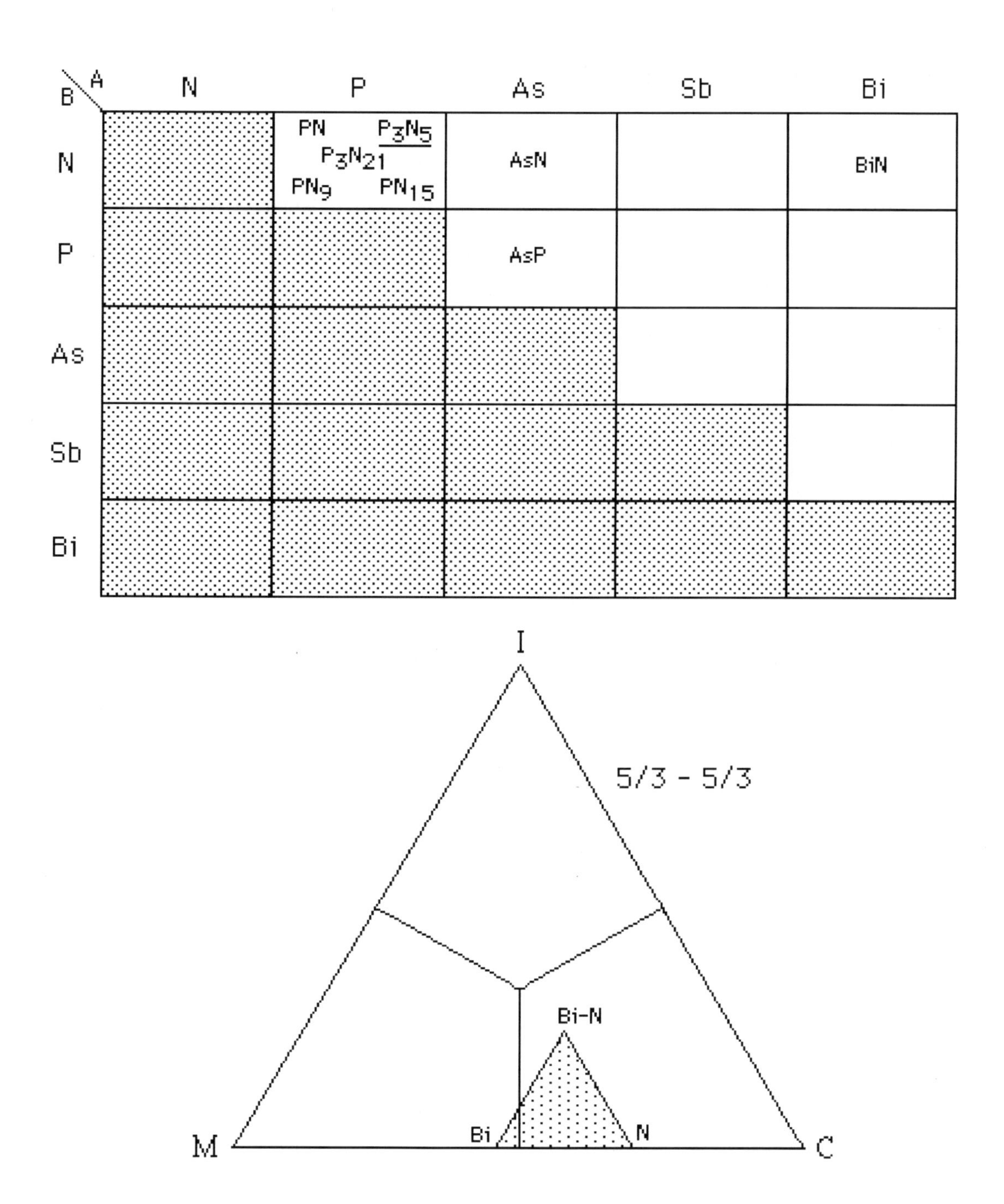

for the N-N bonds to a maximum of 54% for the Bi-Bi bonds or, conversely, the covalency ranges from a maximum of 70% for the nitrogen bonds to a minimum of 46% for the bismuth bonds.

8.5 Group *5/3* - Group *4/4* Binaries

The net stoichiometries of most of the known binary compounds formed between the Group *5/3* elements and carbon, the only member of Group *4/4* more electronegative than the heavier members of Group *5/3*, are summarized in the following table (1, 2):

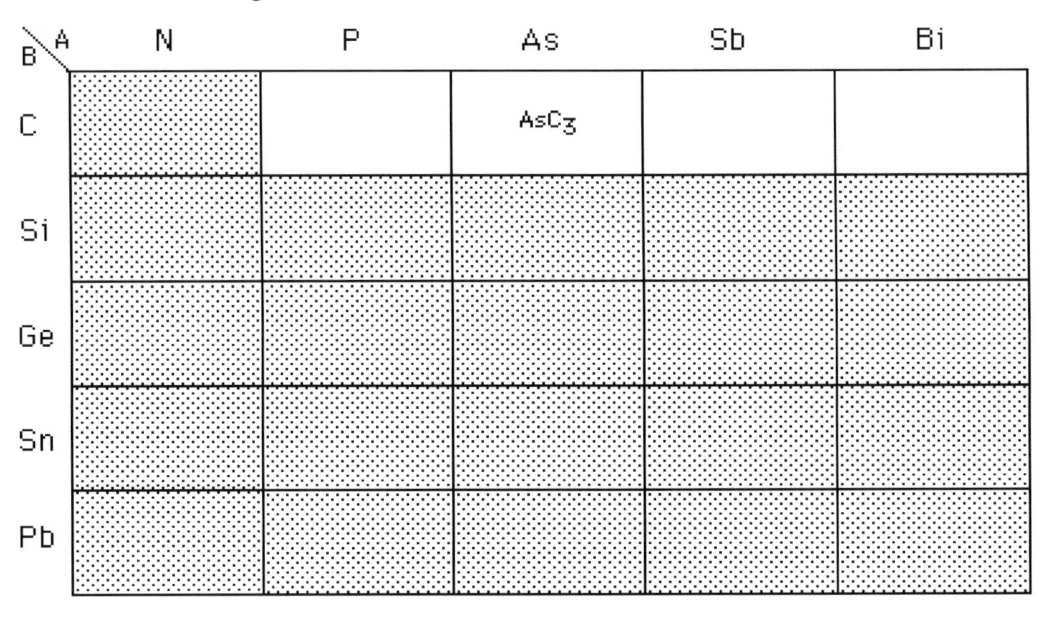

B \ A	N	P	As	Sb	Bi
C			AsC_3		
Si					
Ge					
Sn					
Pb					

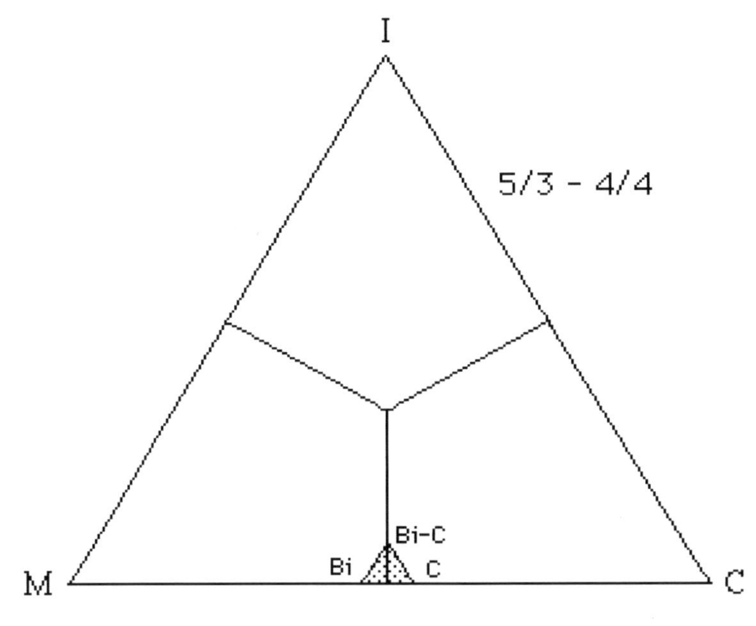

The stoichiometry for homodesmic or simple binaries is predicted to be A_4B_5. Thus for homodesmic binaries: $b/a = e_A/v_B = 5/4$, for heterodesmic polyanionic binaries: $b/a > 5/4$, and for heterodesmic polycationic binaries: $b/a < 5/4$. The only known binary compound appears to be a polycarbide of arsenic, $As(C_3)$. The largely hypothetic parallelogram for the resulting bond-types had been reduced to a small triangle due to its intersection with the base of the triangle and falls almost exactly on the border between the predominantly covalent region and the predominantly metallic region.

8.6 Group *6/2* - Group *7/1* Binaries

The net stoichiometries of most of the known binary compounds formed between the Group *6/2* elements and the halogens or Group *7/1* elements are summarized in the following table (1, 2):

B \ A	O	S	Se	Te
F	O_4F_2 O_2F_2 OF_2	$S_\alpha F_2$ S_2F_2 S_2F_4 S_2F_6 S_4F_8 S_2F_{10} $\underline{SF_6}$	SeF_4 $\underline{SeF_6}$	TeF_4 $\underline{TeF_6}$
Cl		S_aCl_2 S_2Cl_2 SCl_2 SCl_4	Se_2Cl_2 $SeCl_2$ $SeCl_4$	Te_2Cl Te_3Cl_2 $TeCl_2$ $TeCl_4$
Br		S_aBr_2 S_2Br_2 SBr_2	Se_2Br_2 $SeBr_2$ $SeBr_4$	Te_2Br $TeBr_2$ $TeBr_4$
I		S_aI_2 S_2I_2	Se_2I_2	Te_2I TeI TeI_2 TeI_4

The stoichiometry for homodesmic or simple binaries is predicted to be AB_6. Thus for homodesmic binaries: $b/a = e_A/v_B = 6/1$, for heterodesmic polyanionic binaries: $b/a > 6/1$, and for heterodesmic polycationic binaries: $b/a < 6/1$. The only known simple binaries are the fluorides of S, Se, and Te, all of the other compounds falling into the class of polycationic binaries.

Once more the bond parallelogram lies completely within the predominantly covalent region. Since the *FEN* value of O is greater than those of Cl, Br and I, intersection of the bond parallelogram with the base eliminates these bonds, which appear instead in the corresponding bond triangle for the Group *7/1* - Group *6/2* bonds as halogen oxides. The ionic character of the bonds varies between 46% for the Po-F bond and 0% for the A-A bonds lying along the base of the triangle, whereas their metallic character varies between 38% for the iodide bonds and 0% for the fluoride bonds, and their covalent character from 85% for the oxygen bonds to 54% for the Po bonds.

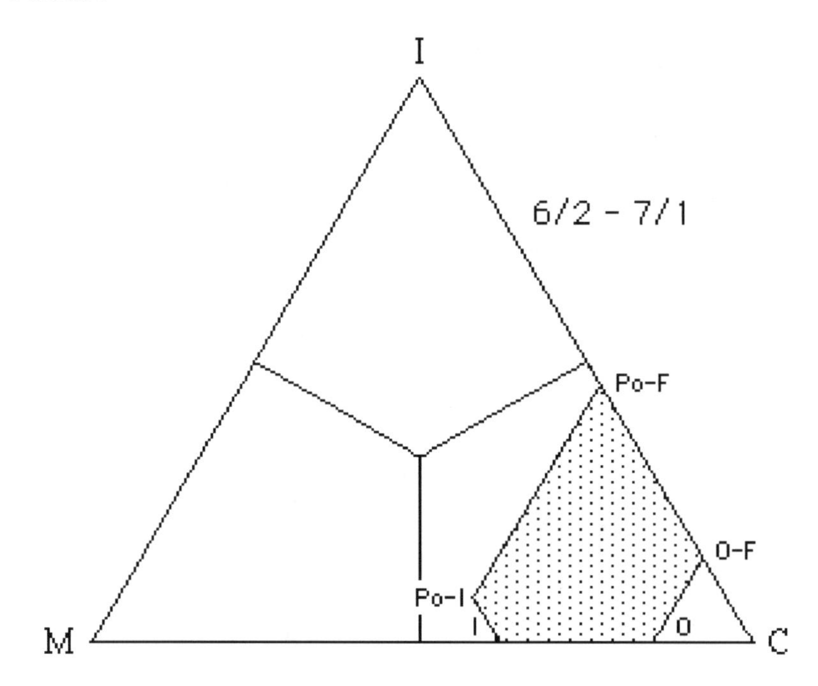

8.7 Group *6/2* - Group *6/2* Binaries

The net stoichiometries of most of the known binary compounds formed among the members of Group *6/2* elements are summarized in the following table (1, 2):

B \ A	O	S	Se	Te
O		S_aO_b (b = 1-2) (a = 2, 6-10) SO_2 $\underline{SO_3}$	SeO_2 Se_2O_5 $\underline{SeO_3}$	TeO_2 Te_2O_5 $\underline{TeO_3}$
S			Se_aS_{6-a} Se_aS_{8-a} Se_aS_{12-a}	
Se				
Te				

The stoichiometry for homodesmic or simple binaries is predicted to be AB_3. Thus for homodesmic binaries: $b/a = e_A/v_B = 6/2 = 3/1$, for heterodesmic polyanionic binaries: $b/a > 3/1$, and for heterodesmic polycationic binaries: $b/a < 3/1$. Only the highest oxides form simple binaries, the remaining oxides being polycationic. The sulfides of Se, on the other hand, are

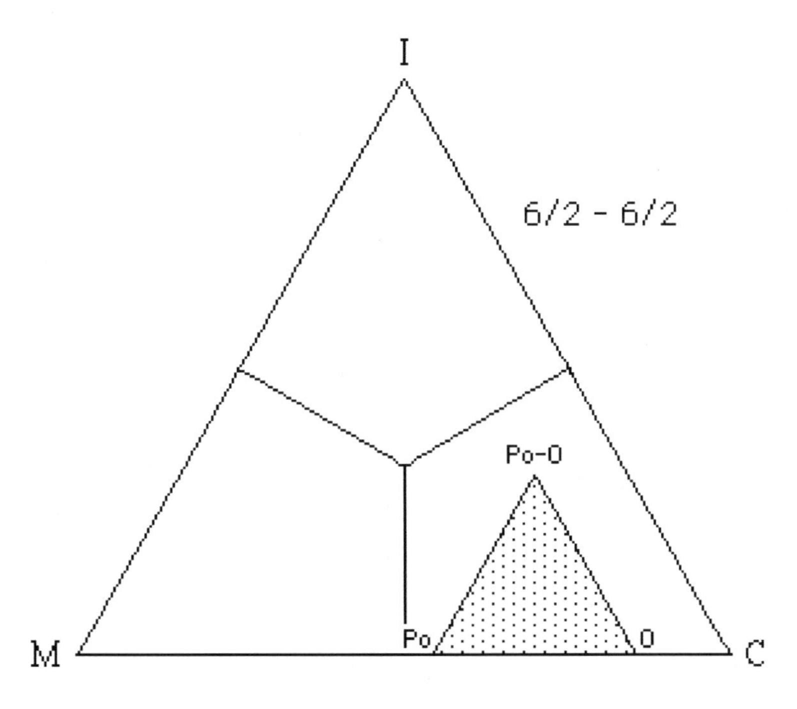

polyanionic and are formed by the progressive substitution of the S atoms for Se within the cyclic molecular anions of the form S_6, S_8 and S_{12}.

The bond parallelogram has been truncated once again to a triangle due to its intersection with the bottom edge of the bond-type triangle, thereby automatically eliminating redundant but inverted bond combinations. The ionic character of the bonds ranges from a maximum of 31% for the Po-O bond to a minimum of 0% for all of the self A-A bonds lying along the bottom edge of the triangle, whereas the metallic character varies from a maximum of 46% for the polonide bonds to a minimum of 15% for the oxide bonds. Likewise the covalent character of the simple substances lying along the base varies from 85% for the O-O bonds to 46% for the Po-Po bonds.

8.8 Group *6/2* - Group *5/3* Binaries

The net stoichiometries of most of the known binary compounds between the Group *6/2* elements and nitrogen, the only member of Group *5/3* more electronegative than the heavier members of Group *6/2*, are summarized in the following table (1, 2):

B \ A	O	S	Se	Te
N		S_aN_2 S_4N_2 S_aN_a ($a = 1, 2, 4, \infty$) S_5N_6 $\underline{SN_2}$	Se_4N_2 Se_4N_4	Te_4N_4
P				
As				
Sb				
Bi				

The stoichiometry for homodesmic or simple binaries is predicted to be AB_2. Thus for homodesmic binaries: $b/a = e_A/v_B = 6/3 = 2/1$ for heterodesmic polyanionic binaries: $b/a > 2/1$, and for heterodesmic polycationic binaries: $b/a < 2/1$. As may be seen, the only known simple binary is SN_2.

Only the upper tip of the bond parallelogram is preserved, with the ionicity ranging from a maximum of 16% for the Po-N bond to 0% for the A-A bond lying along the bottom edge of the triangle. The metallicity of these A-A bonds varies from 45% for the Po-Po bond to 30% for the N-N bond, and the covalency from 70% for the N-N bond to 54% for the Po-Po bond.

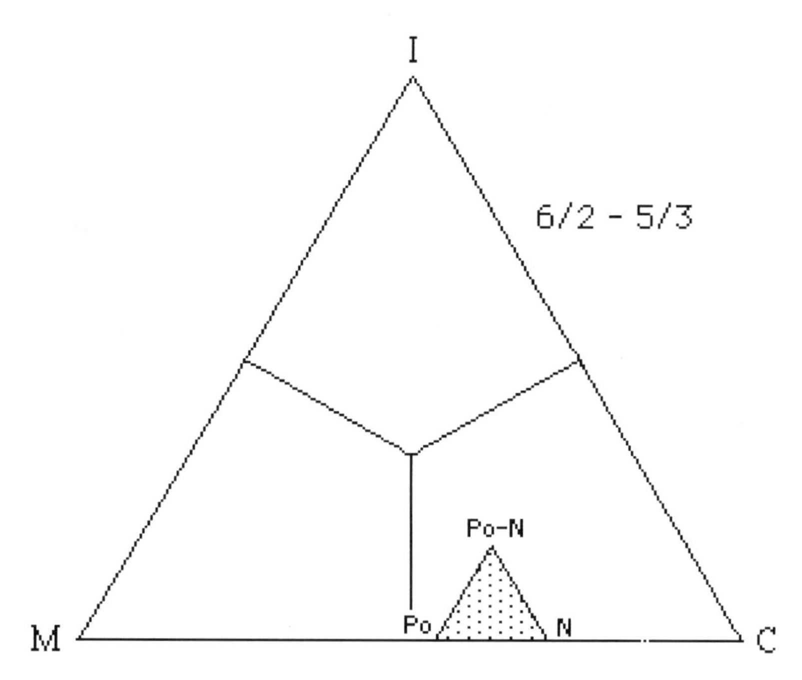

8.9 Group *7/1* - Group *7/1* Binaries

The net stoichiometries of most of the known binary compounds formed among the members of the Group *7/1* elements are summarized in the following table (1, 2):

B \ A	F	Cl	Br	I
F		ClF ClF_3 ClF_5	BrF BeF_3 BrF_5	IF IF_3 IF_5 $\underline{IF_7}$
Cl			BrCl	ICl I_2Cl_6
Br				IBr
I				

The stoichiometry for homodesmic or simple binaries is predicted to be AB_7. Thus for homodesmic binaries: $b/a = e_A/v_B = 7/1$, for heterodesmic polyanionic binaries: $b/a > 7/1$, and for heterodesmic polycationic binaries: $b/a < 7/1$. The only known simple binary is IF_7. All of the other compounds fall into the polycationic category and involve cationic lone pairs rather than A-A bonding.

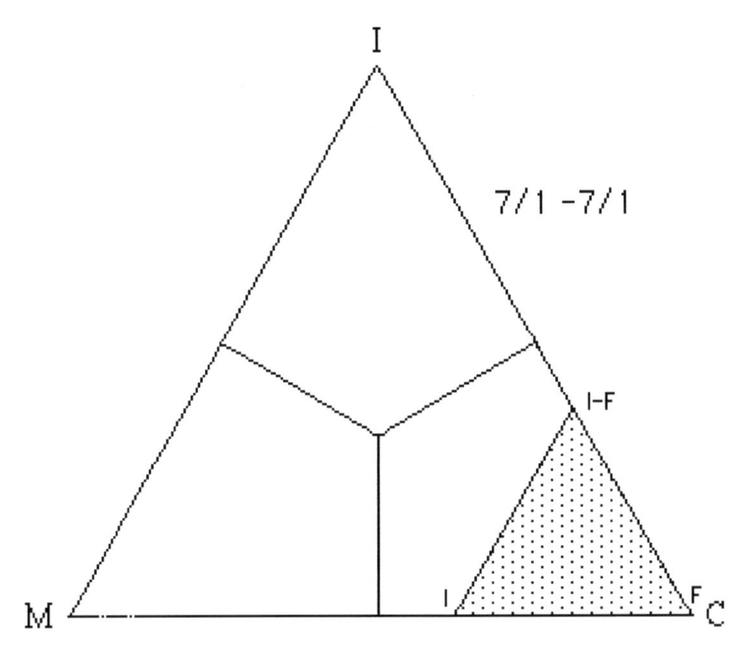

The resulting truncated bond parallelogram lies completely within the predominantly covalent region, with the ionicity varying from a maximum of 38% for the I-F bond to a minimum of 0% for the A-A bonds lying along the base of the triangle. The metallicity of these A-A bonds varies from a maximum of 38% for the I-I bond to a minimum of 0% by definition for the F-F bond, and the covalency from a maximum of 100% by definition for the F-F bond to a minimum of 62% for the I-I bond. Note that the unusual widths of the bond parallelograms for Groups *5/3-8/0* reflect the increasing intragroup spread in the *FEN* values as one moves toward the right side of the periodic table rather than an increase in the number of distinct bonds.

Given the size of the truncated parallelogram in this bond-type triangle versus the number of compounds listed in the previous table, it is perhaps worth reminding the reader, as was detailed in Section 5.3, that the boundaries of the parallelograms in these diagrams, whether complete or truncated, are determined by the spread in the electronegativity values for the bond components under consideration and that, consequently, their resulting areas in no way reflect the relative density of data points contained within them.

8.10 Group *7/1* - Group *6/2* Binaries

The net stoichiometries of most of the known binary compounds between the Group *7/1* elements and oxygen, the only member of Group *6/2* more electronegative than the heavier members of Group *7/1*, are summarized in the table at the top of the following page (1). The stoichiometry for homodesmic or simple binaries is predicted to be A_2B_7. Thus for homodesmic binaries: $b/a = e_A/v_B = 7/2$, for heterodesmic polyanionic binaries: $b/a > 7/2$, and for heterodesmic polycationic binaries: $b/a < 7/2$. The only known simple binary is Cl_2O_7. All of the other compounds fall into the polycationic category and involve cationic lone pairs rather than A-A bonding.

The tip the truncated bond parallelogram lies completely within the predominantly covalent region, with the ionicity varying from a maximum of

B \ A	F	Cl	Br	I
O		Cl_2O Cl_2O_3 Cl_2O_4 ClO_2 Cl_2O_6 $\underline{Cl_2O_7}$	Br_2O Br_2O_3 Br_2O_5	I_2O_4 I_2O_5 I_2O_6
S				
Se				
Te				

23% for the I-O bond to a minimum of 0% for the A-A bonds lying along the base of the triangle. The metallicity of these A-A bonds varies from a maximum of 38% for the I-I bond to a minimum of 15% for the O-O bond, and the covalency from a maximum of 85% for the O-O bond to a minimum of 62% for the I-I bond

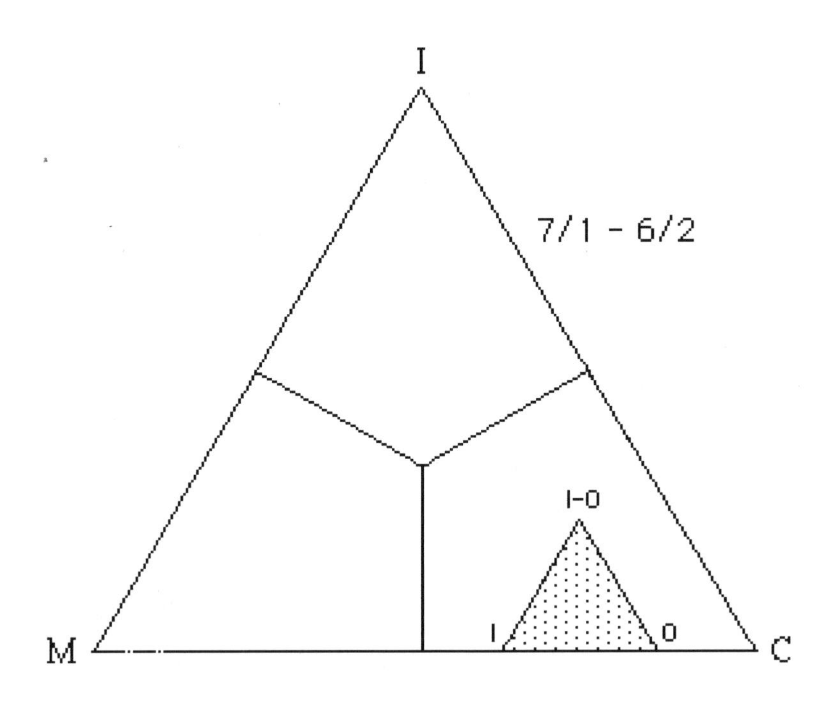

8.11 Group *7/1* - Group *5/3* Binaries

The net stoichiometries of most of the known binary compounds between the Group *7/1* elements and nitrogen, the only member of Group *5/3* more electronegative than the heavier members of Group *7/1*, are summarized in the following table (1):

B \ A	F	Cl	Br	I
N			Br_3N BrN_3	I_3N IN_3
P				
As				
Sb				
Bi				

The stoichiometry for the homodesmic or simple nitrides is predicted to be A_3B_7. Thus for homodesmic binaries: $b/a = e_A/v_B = 7/3$, for heterodesmic polyanionic binaries: $b/a > 7/3$, and for heterodesmic polycationic binaries: $b/a < 7/3$. No simple binaries are known, all of the listed compounds being instead either polycationic (A_3B) or polyanionic (AB_3). It is of interest to note that when one consistently applies the convention of writing the more electropositive component of a compound first, then the explosive compound nitrogen triiodide (NI_3), much beloved of Freshman chemistry demonstrations, should, for consistency, actually be written as triodine nitride (I_3N). Indeed, the actual material used in the demonstration is not the pure binary compound, but rather an adduct containing an additional ammonia ligand ($I_3N{\bullet}NH_3$) (3, 4). Also note that, according to electronegativity conventions, BrN_3 and IN_3 are both formally polyanionic azides.

The tip of the truncated bond parallelogram lies completely within the predominantly covalent region, with the ionicity varying from a maximum of 8% for the I-N bond to a minimum of 0% for the A-A bonds lying along the base of the triangle. The metallicity of these A-A bonds varies from a maximum of 38% for the I-I bond to a minimum of 30% for the N-N bond, and the covalency from a maximum of 70% for the N-N bond to a minimum of 62% for the I-I bond.

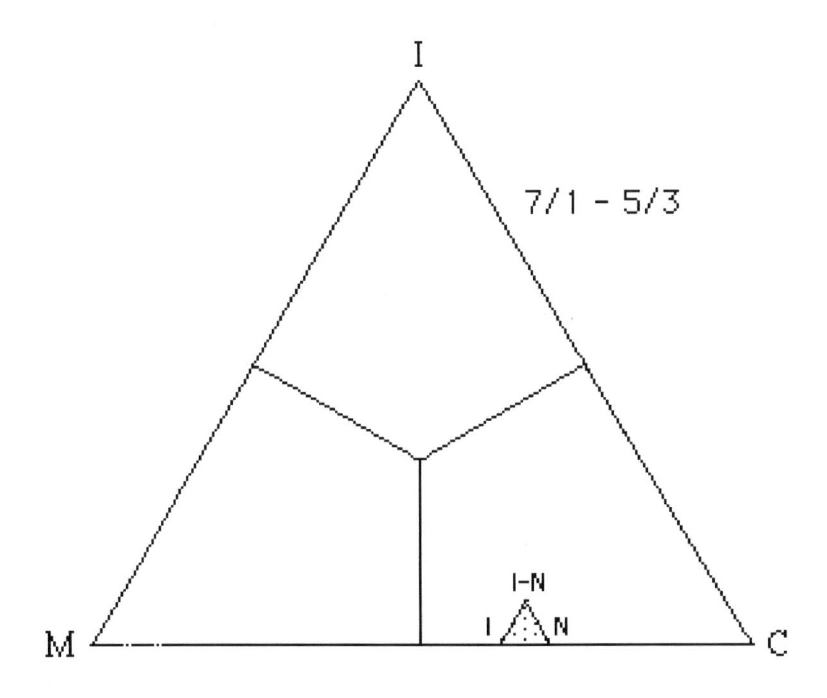

8.12 Group *8/0* - Group *7/1* Binaries

The net stoichiometries of the known binary compounds formed between the Group *8/0* elements and fluorine and chlorine, the only members of Group *7/1* that are more electronegative than the heavier members of Group *8/0*, are summarized in the following table (1):

B \ A	Ne	Ar	Kr	Xe
F			KrF_2	XeF_2 XeF_4 XeF_6
Cl				
Br				
I				

The stoichiometry for homodesmic or simple binaries is predicted to be AB_8. Thus for homodesmic binaries: $b/a = e_A/v_B = 8/1$, for heterodesmic polyanionic binaries: $b/a > 8/1$, and for heterodesmic polycationic binaries: $b/a < 8/1$. As may be seen, no examples of simple binaries are known. All of the known compounds are instead polycationic and involve cationic lone pairs

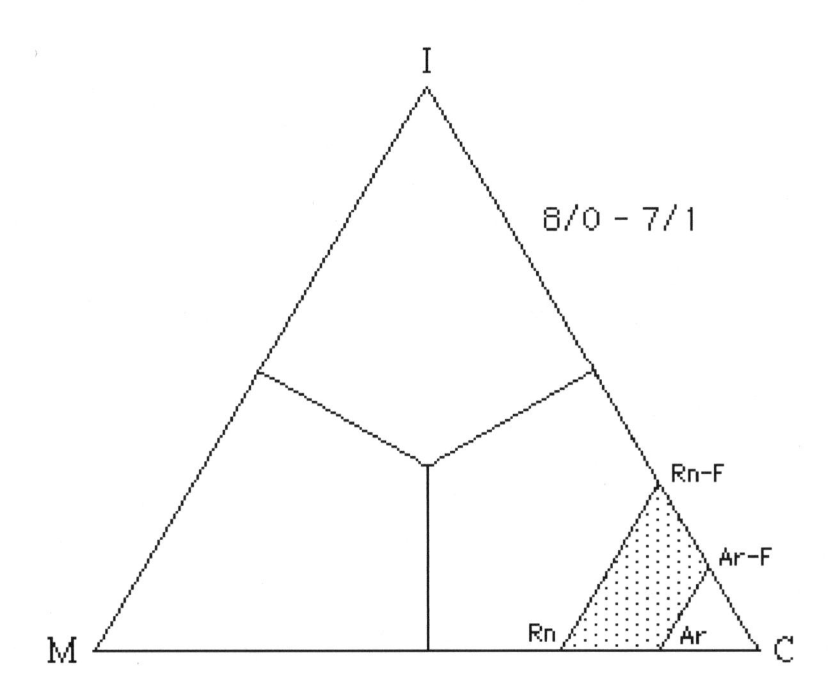

rather than A-A bonding.

The truncated bond parallelogram lies within the predominantly covalent region and is relatively narrow due to the fact that Ne has been excluded since its *FEN* value is greater than that of F and, unlike the heavier noble gases, it appears to be truly chemically inert. The ionicity varies from a maximum of 30% for the Rn-F bond to a minimum of 0% for the free A atoms lying along the base of the triangle and the covalency along the base varies from a maximum of 85% for Ar to a minimum of 70% for Rn.

8.13 Group *8/0* - Group *6/2* Binaries

The net stoichiometries of the known binary compounds formed between the Group *8/0* elements and oxygen, the only member of Group *6/2* that is more electronegative than the heavier members of Group *8/0*, are summarized in the following table (1):

B \ A	Ne	Ar	Kr	Xe
O				XeO_3 $\underline{XeO_4}$
S				
Se				
Te				

The stoichiometry for homodesmic or simple binaries is predicted to be AB_4. Thus for homodesmic binaries: $b/a = e_A/v_B = 8/2 = 4/1$, for heterodesmic polyanionic binaries: $b/a > 4/1$, and for heterodesmic polycationic binaries: $b/a < 4/1$. As may be seen, only known simple binary is XeO_4. The only known polycationic binary, XeO_3, exhibits cationic lone pairs rather than A-A bonding.

The tip of the resulting truncated bond parallelogram lies in the predominantly covalent region, with the ionicity varying from a maximum of 15% for the Rn-O bond to a minimum of 0% for the free A atoms lying along the base of the triangle.

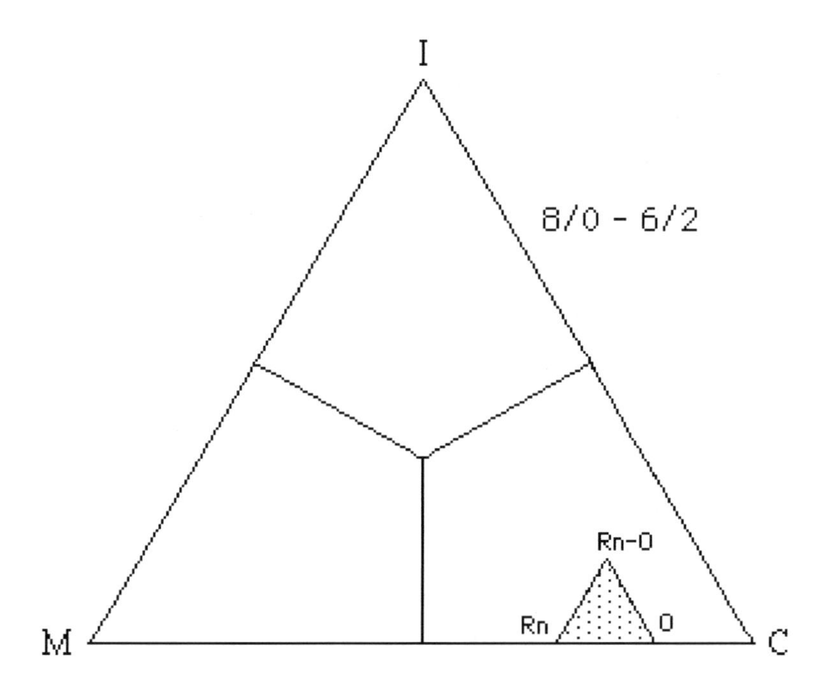

8.14 References and Notes

1. N. Wiberg, *Inorganic Chemistry*, Academic Press: New York, NY, 2001, Chapters XI-XIV.

2. B. Predel, Ed., *Phase Equilibria, Crystallographic and Thermodynamic Data of Binary Alloys*, Landolt-Börnstein, Vols. 5a-5i Springer: Berlin, 1991-1998.

3. N. N. Greenwood, A. Earnshaw, *Chemistry of the Elements*, Pergamon: Oxford, 1994, p. 506.

4. T. Mak, G-D. Zhou, *Crystallography in Modern Chemistry,* Wiley: New York, NY, 1992, pp. 346-349.

IX
Empirical Sorting Maps

9.1 The Nature of Sorting Maps

Since the time of Galileo and Newton the archetypical model of science, much beloved of philosophers, has centered around the use of theoretical models to deduce explicit mathematical relationships between quantifiably measurable properties. In other words, it has centered around making explicit mathematical functions of the form:

$$u = f(x, y, z \ldots) \qquad [1]$$

where u, x, y, and z are properties with assignable numerical values. Even in the absence of a fundamental theory relating the properties in question, it is possible, if a sufficiently broad range of property values have been experimentally measured, to mimic this process by empirically fitting the values to some functional curve – a technique which forms the basis of a great deal of applied engineering.

This particular model of science is heavily biased toward the science of physics and soon proves inadequate when one considers more qualitative sciences, such as biology, which instead rely heavily on the use of qualitative, descriptive classification. Chemistry forms an unique bridge between these two extremes. Like physics, it makes use of many quantitative mathematical functions or laws, but like biology, it also employs many qualitative class concepts, such as acid, base, element, compound, reducing agent, oxidizing agent, halogen, aldehyde, alcohol, etc.

Unlike the causal relationship typified by relation 1, the problem of describing class concepts can seldom be given an explicit functional form. In the simplest possible case, the class behavior will correlate with certain ranges of a quantifiable property x of the species in question:

$$Class\ A\ =\ f(range\ A\ of\ x) \quad [2]$$

$$Class\ B\ =\ f(range\ B\ of\ x),\ etc. \quad [3]$$

That is, instead of the simple one to one mapping characteristic of relation 1, we will have a many to one mapping.

However, since recourse to the use of class concepts in the first place is frequently a reflection of the complexity of the phenomenon being described, the most common case is usually a good deal more complex:

$$Class\ A\ =\ f(range\ A\ of\ x,\ range\ A\ of\ y,\ range\ A\ of\ z\ ...) \quad [4]$$

where, although a species may have a value of x lying within the required range for class A, it will still not correlate with that class unless it simultaneously possesses values of y and z also falling within the required ranges.

Moreover, the components of relations 2-4 may vary widely in their precision. Classes may correspond to easily recognizable either/or situations, such as whether a series of solids has structure A versus structure B, or whether binary combinations of liquids are or are not completely miscible with one another, or they may correspond to some ill-defined but widely used distinction, such as whether the species in question are or are not metallic. Likewise, the properties correlating with the class behavior may range from those with assignable numerical values, such as electronegativity, radius, or ionic charge, to more vague concepts, such as the ability to undergo a certain type of reaction.

A useful empirical procedure for making relationships 2-4 more visually explicit is a device known as a sorting map (1). This requires a set of species, (called the learning set) whose class assignments or behavior are already known, as well as a set of quantified properties (known as the sorting parameters) whose numerical values are thought to determine or correlate with the class behavior in question. Using these properties as the coordinates, the learning set is then plotted using distinctive point symbols for the members of each class. The result is not a straight line or a curve of some sort, as in the classic situation given by function 1, but rather a scattering of the points throughout the plane of the paper. However, if successful, this scattering is not random, but rather results in the data points for each class being concentrated in separate, nonoverlapping regions of the plane. Such success implies that the properties used as the sorting parameters play a significant role in determining class behavior, and the further addition of best-guess class boundaries to the sorting map then allows one to interpolatively predict the class behavior of yet other species not included in the original learning set.

One of the earliest successful sorting maps was constructed by the Swiss-Norwegian geochemist, Victor Goldschmidt, in 1934 in order to characterize the chemical behavior of ions with respect to water (2). Using the ionic charge (*x*-axis) and the ionic radius (*y*-axis) as the sorting parameters, he was able to separate the ions into three classes: those which formed stable, hydrated ions and were localized in the upper left-hand corner of his diagram, those which reacted with water to give various insoluble hydroxide and oxide precipitates and were localized in the center of his diagram, and those which gave complex, water-soluble oxo anions, such as CO_3^{2-} and SO_4^{2-}, and were localized in the lower right-hand corner of his diagram. Since then, sorting maps have been proposed for a wide variety of chemical purposes, including the prediction of metal-metal solubilities (3), the sorting of ions into the classes of hard and soft Lewis acids and bases (4), and the prediction of compound formation in binary phase diagrams (5), to name but a few.

9.2 Structure

In 1959 Mooser and Pearson published the first structure-sorting map for main-block, homodesmic compounds of stoichiometry AB using the electronegativity difference ($\Delta EN = EN_B - EN_A$) of the component atoms and the average principal quantum number of their valence shells ($n_{av} = (n_A + n_B)/2$) as the sorting parameters (6):

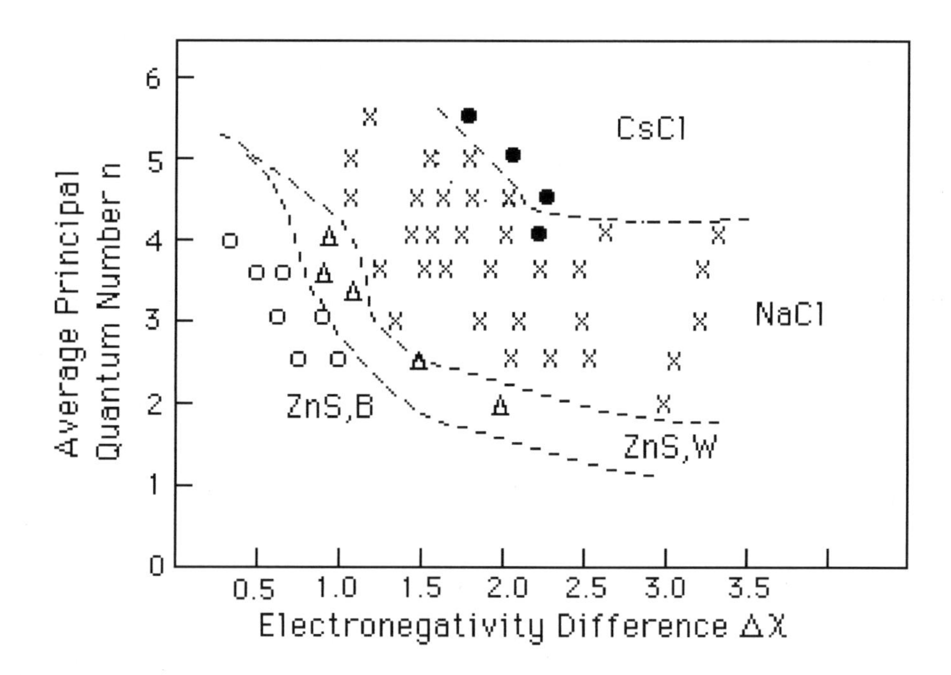

As may be seen, the resulting sorting map successfully separated these compounds into distinctive regions corresponding to those having the 8/8 CsCl framework structure, the 6/6 NaCl framework structure, and the two polymorphic forms (wurtizite and zinc blende) of the 4/4 ZnS framework structure, where the numerical ratios represent coordination number ratios (CN_B/CN_A) rather than e_A/v_B ratios (7). A successful structural sorting of these compounds was also obtained by Goryunova in 1963 using the ratio of the core electron affinity to the core charge ($N' = I_Z/Z$) of each component atom as the sorting parameters (8), and in 1969 Phillips published a similar result using a set of theoretical sorting parameters related to the ionic (*iC)* and covalent (E_h) contributions to the band-gap energy (*Eg*) of each species (9).

In 1974 Muller and Roy showed that similar structure-sorting maps could be constructed for various families of ternary compounds of composition $A_aB_bC_c$ using the empirical ionic radii of A and B as the sorting parameters, provided that both the overall compound stoichiometry and the nature of C were kept constant for each sorting map (10) – a result which was later duplicated by many others over the next decade using theoretically calculated pseudopotential "orbital" radii instead (11).

The author's personal interest in bond-type triangles was actually a by-product of his attempt to improve the original structure-sorting map of Mooser and Pearson (12). These authors had noted that the ΔEN parameter could be identified with compound ionicity and the n_{av} parameter with compound metallicity. However, while *n* remained constant for each period of the main-block, the metallicity of the corresponding simple substances quite obviously did not. Consequently it occurred to the author that the parameter $1 - EN_{av}$ might provide a measure of metallicity much more in keeping with observed group and period trends. Use of ΔEN and $1 - EN_{av}$ as the sorting parameters not only produced a successful structure-sorting map for homodesmic AB compounds and simple substances, it also allowed one to separate the resulting structure regions using straight lines instead of the arbitrary curves used by Mooser and Pearson.

In addition, it was soon apparent that the resulting data points were also superimposable on a triangular array similar to the well-known qualitative van Arkel version of the bond-type triangle and that was possible to replace the original metallicity parameter with EN_{av} alone, which appeared instead to function as a so-called covalency parameter. Though used for several years in the author's chemistry lectures, it was not until 1995 that this particular form of the combined AB structure-sorting map and bond-type triangle was finally published in response to the related work of Allen and Sproul, where the numbers in the various sorting regions indicate over-lapping data points (13):

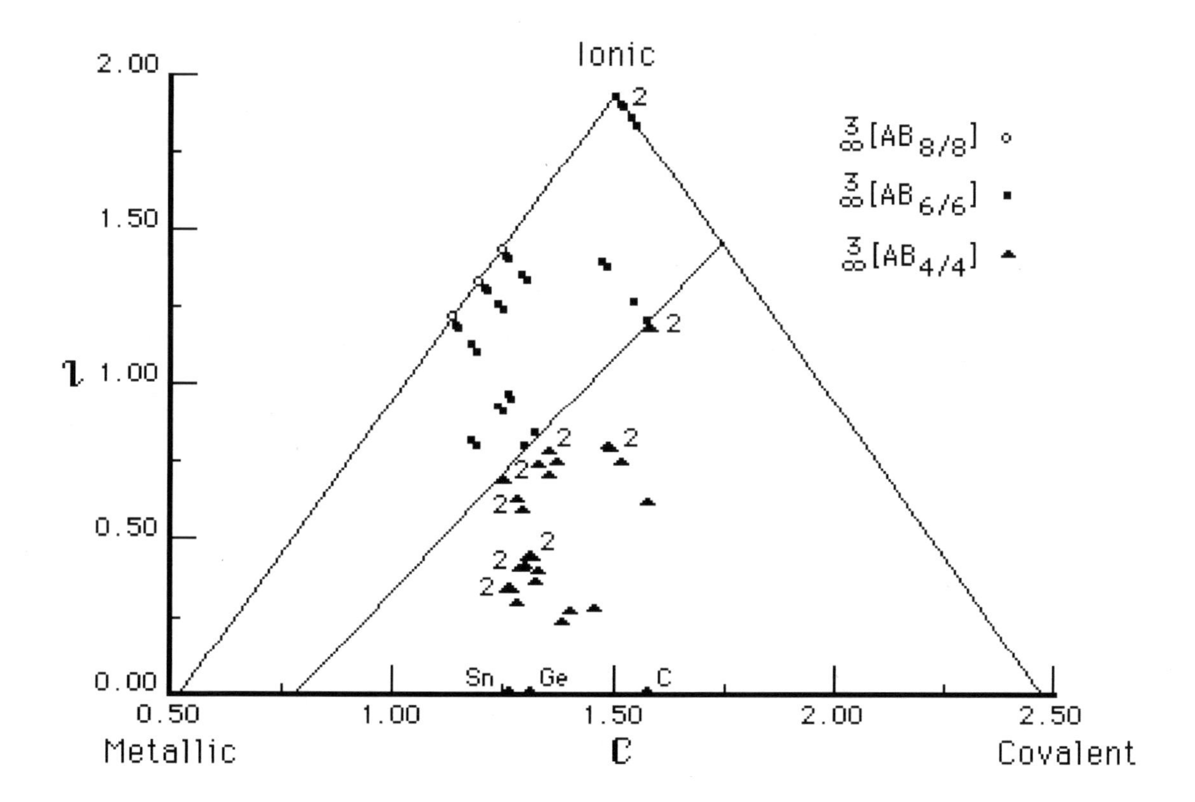

Recall that we obtain the identical result if we use *FEN* values rather than *EN* values.

If we return to the general function discussed in Section 4.6:

$$Properties = f(FEN_A, FEN_B, b/a, m, P, T) \qquad [5]$$

it is apparent that the above result corresponds to the special case:

$$Structure = f(FEN_A, FEN_B)_{b/a,m,P,T} \qquad [6]$$

in which $b/a = e_A/v_B = 1/1$, m = main-block, and P and T correspond to room temperature and pressure (RTP).

This is made more apparent by replotting the AB data, which is summarized in Appendix II, using the Yeh orientation of the Grimm bond-type triangle. This has been done in the following three figures in which each structure class is plotted separately (with respect, of course, to the identical boundary lines) in order to eliminate some of the clutter of the original plot:

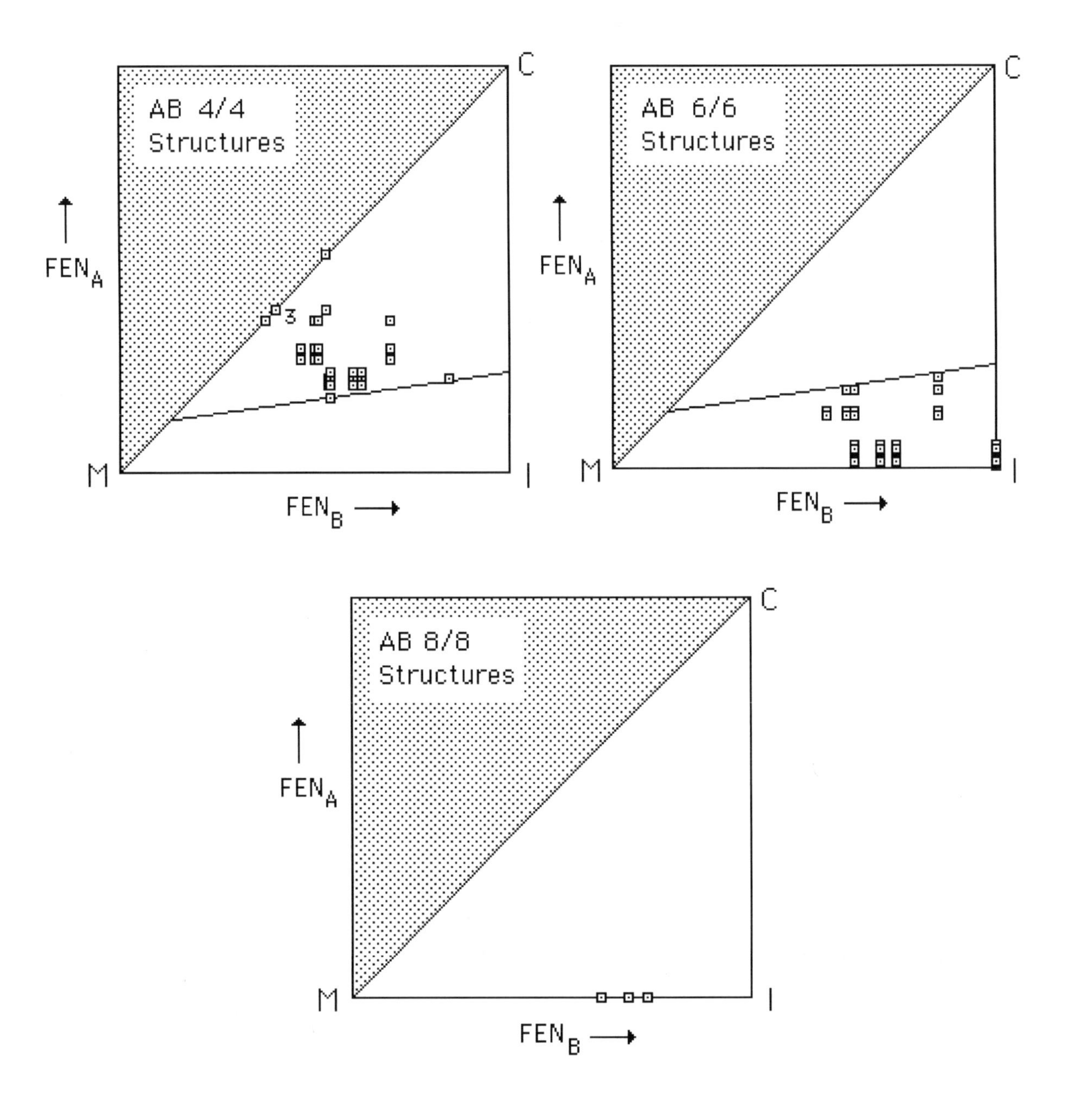

Though this makes it easier to visually assess the level of class separation, it is unable to correct the additional problem that several of the points overlap and cannot be resolved at the level of magnification used here.

Yet a third method of presenting this material is based on the following table, which summarizes the predicted stoichiometries of the various main-block, homodesmic, binary compounds discussed in the previous four lectures:

B \ A	1/7	2/6	3/5	4/4	5/3	6/2	7/1	8/0
1/7	A_7B							
2/6	A_6B	A_3B						
3/5	A_5B	A_5B_2	A_5B_3					
4/4	A_4B	A_2B	A_4B_3	AB	A_4B_5			
5/3	A_3B	A_3B_2	AB	A_3B_4	A_3B_5	AB_2	A_3B_7	
6/2	A_2B	AB	A_2B_3	AB_2	A_2B_5	AB_3	A_2B_7	AB_4
7/1	AB	AB_2	AB_3	AB_4	AB_5	AB_6	AB_7	AB_8

Examination of this table shows that all of the AB compounds in our structure sorting map result from just four intergroup combinations: Group *1/7* - Group *7/1*, Group *2/6* - Group *6/2*, Group *3/5* - Group *5/3* and Group *4/4* - Group *4/4*. The corresponding bond parallelograms for each of these four combinations are shown on the bond-type triangle at the top of the following page and superimposed on each are the boundaries for our three structure classes.

As may be seen, these four parallelograms form a diagonal progression which begins in the predominantly ionic region along the upper left-hand edge of the triangle and ends in the intersection of the last parallelogram with of the base of the triangle just to the left of the line separating the predominately metallic and predominantly covalent regions.

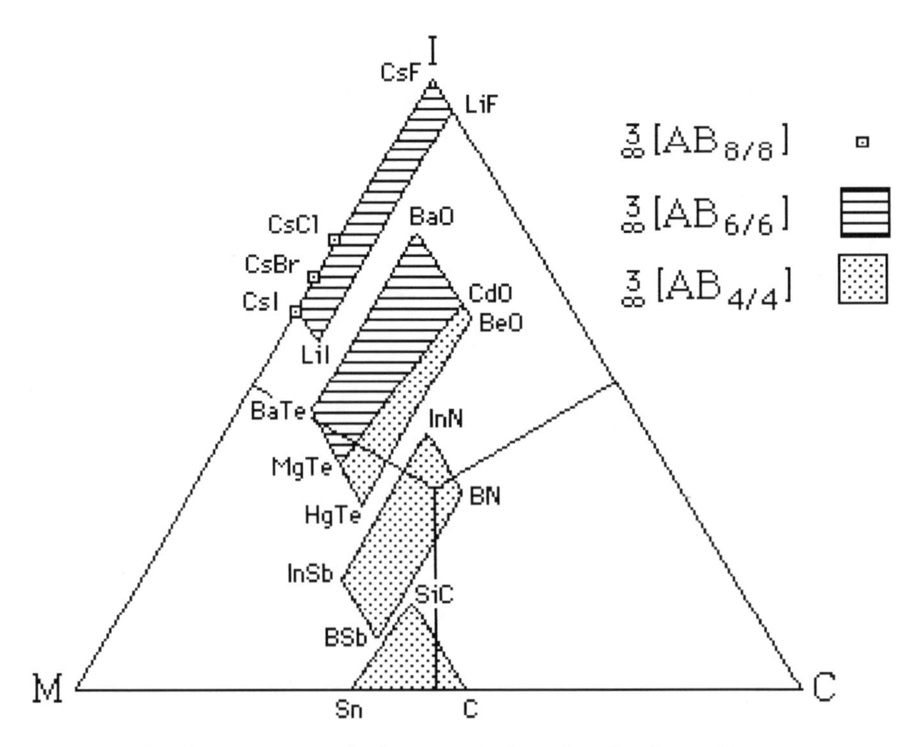

This approach does a much better job of relating the structure sorting function to the bond-type function and clearly demonstrates the validity of Jørgensen's warning many years ago concerning the questionable validity of the tacit assumption of Wells and others that a 4/4 ZnS framework structure automatically indicates covalent bonding whereas a 6/6 NaCl framework structure automatically indicates ionic bonding (14).

As summarized in the table at the top of the following page, which is based on the above comprehensive table, similar structure-sorting maps may be constructed for other isostoichiometric classes of binary, homodesmic compounds. However, none of them will be as extensive or as varied as the sorting map for the AB compounds.

9.3 Band-Gap Energies

The initial development AB structure-sorting maps was a by-product of an intense interest in the 1950s, 1960s and 1970s in the potential semiconducting

Inter- and Intragroup Combinations Leading to Identical Stoichiometries

Stoichiometry	*Group Combinations*
AB	1/7-7/1, 2/6-6/2, 3/5-5/3, 4/4-4/4
AB_2	2/6-7/1, 4/4-6/2, 6/2-5/3
AB_3	3/5-7/1, 6/2-6/2
AB_4	4/4-7/1, 8/0-6/2
AB_5	5/3-7/1
AB_6	6/2-7/1

properties of these compounds (6, 8-9, 15-17). This suggests that it might not be unreasonable to also search for an approximate sorting of AB homodesmic compounds into the broad classes of insulators and semi-conductors on the basis of both their band-gap energies and bond types, as expressed by the approximate function:

$$Band\ Gap\ =\ f(FEN_A, FEN_B)_{b/a,m,P,T} \qquad [7]$$

However, while structure is an unambiguous, discrete, either/or, property (provided that one ignores the complication of metastable polymorphs), band-gap energies vary in a continuous fashion and a division AB compounds into the either/or classes of insulators versus semiconductors requires that we artificially impose numerical cutoffs in order to define each class. Adopting the pronouncement of Rose *et al* that semiconductors "are characterized by a nearly full valence band which is separated from a nearly empty conduction band by an energy gap of 2 *eV* or less," and using the data summarized in Appendix II, we have constructed the following two sorting maps using the Yeh orientation of the Grimm bond-type triangle (17):

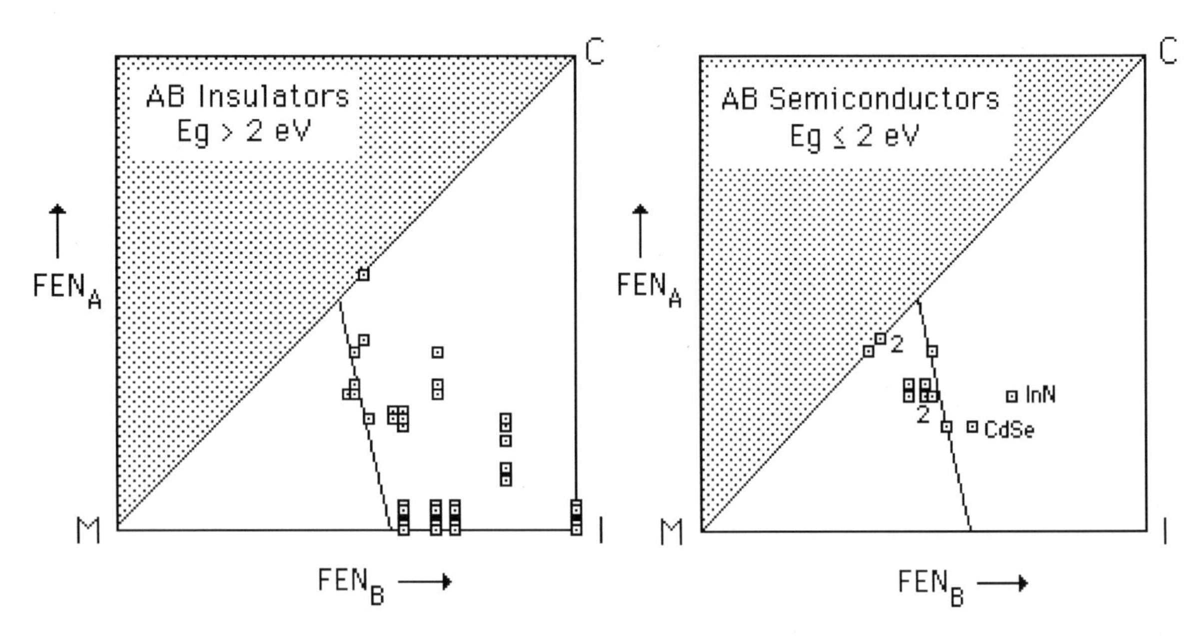

As before, we have shown each class on a separate triangle (maintaining, of course, the identical boundary line) in order to eliminate clutter and to allow for easier visual assessment of the level of class separation, though once again

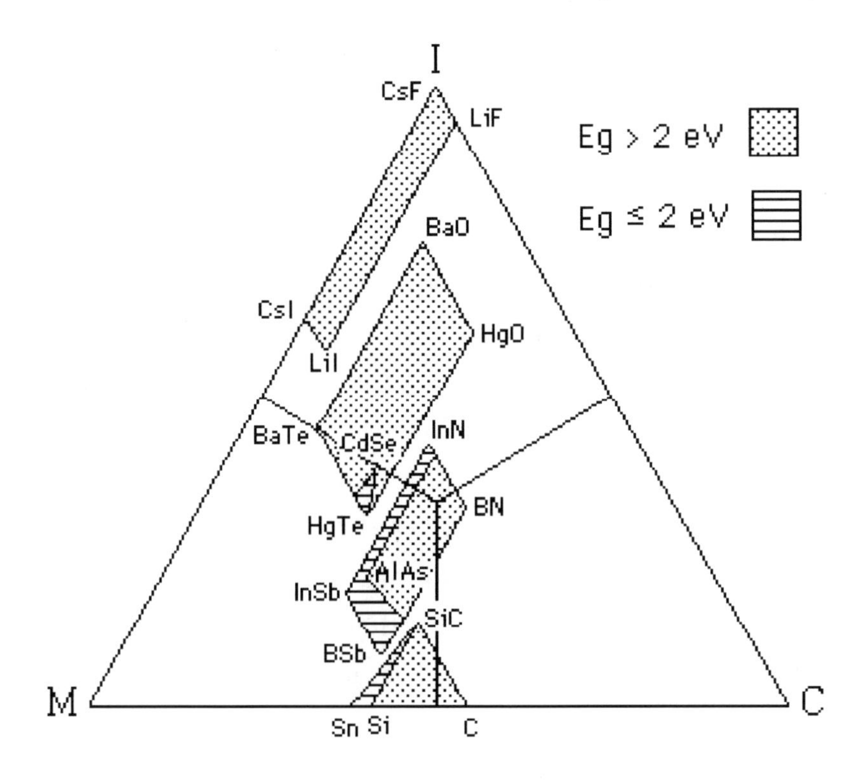

there are problems with poor resolution of partially overlapping data points. While the level of separation is not quite as impressive as in the case of the structure-sorting maps (the most glaring exceptions being the data points for CdSe and InN), the degree of correlation is nevertheless significant.

Replotting the same data on the Fernelius triangle shown on the previous page and superposition of similar boundaries (this time redrawn to accommodate the exceptions) on the corresponding bond parallelograms reveals that, with the exception of InN, all of the semiconducting species fall within the predominantly metallic region of the bond-type triangle suggesting that this is a necessary, though by no means sufficient, condition for significant semiconductivity since some insulators also fall into this region.

9.4 Color

Building on a 1918 paper by Bichowski, Pitzer and Hildebrand suggested in 1941 that there should be an approximate correlation between color and bond covalency – a correlation also predicted by the ionic-polarization model (18, 19). In terms of modern theory, we would anticipate that an approximate correlation would exist between the bond characters of those homodesmic, binary, main-block, compounds under consideration and their colors provided that these colors are largely the result of either B→A charge-transfer transitions or band-gap transitions. Exclusion of the transition elements automatically precludes color generation by means of d-d transitions and the further requirement that the bonding must be homodesmic largely eliminates the possibility of color generation by means conjugated-pi transitions (20, 21):

$$\textit{Charge-Transfer Color} = f(FEN_A, FEN_B)_{b/a,m,P,T} \qquad [8]$$

Once again the necessary data is summarized in Appendix II and we have shown each class on a separate Yeh triangle (maintaining, as always, the identical boundary line) in order to eliminate clutter and to allow for easier visual assessment of the level of class separation:

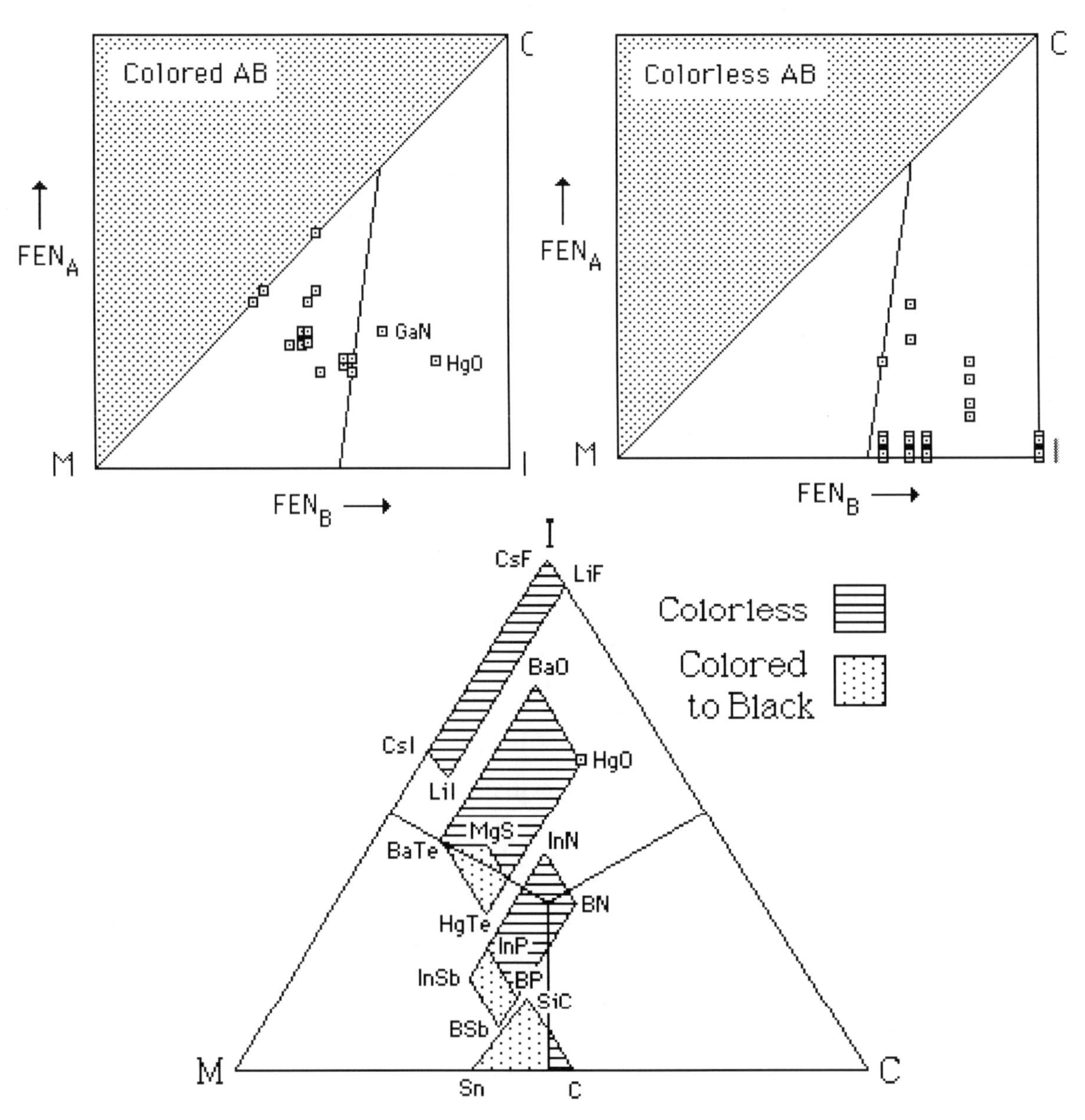

As was the case with the insulator-semiconductor sorting map, the degree of separation is not quite as impressive as was the case with the structure-sorting maps (the most glaring exceptions being the points for GaN and HgO). Nevertheless the qualitative trend is significant and easily visualizable when superimposed on the four bond parallelograms in the corresponding Fernelius triangle, which clearly reveal that it is the development of bond metallicity rather than bond covalency, as originally assumed by Pitzer and Hildebrand, which correlates most strongly with the development of color in these compounds.

It should be further noted that the data set for this plot is relatively soft, being based on the colors reported in the *Handbook of Chemistry and Physics* (22). Since these colors are usually mentioned only incidentally in the course of papers dealing with the other concerns, they are often not very accurately characterized. In addition, even slight contamination with excess reactants or lower polyanionic by-products (e.g., polysulfides, polyselenides, etc.) can easily result in a false attribution of color to the dominant homodesmic species.

9.5 Bulk Conductivity

One very crude use of the sorting map concept in conjunction with the bond-type triangle is to superimpose the results of a Freshman chemistry demonstration designed to illustrate the relationship between bond type and bulk electrical conductivity for simple substances and binary compounds. This employs a simple "either/or" conductivity tester and a series of solid and liquid samples of varying bond character. As shown on the following page, the either/or conductivity tester is made by connecting a 1.5-volt dry cell to an electric buzzer with a break in the circuit formed by two conductivity probes separated by a one-centimeter gap. When the probes are touched to the solid samples or immersed in the liquid samples, either sufficient current traverses the gap to set off the buzzer or it doesn't.

When the A-B bonds for the samples in question are plotted on the bond-type triangle, those giving a positive conductivity result all fall within the shaded or predominantly metallic region of the triangle, whereas those giving a negative result all fall within the unshaded or predominantly ionic and covalent regions. Of the two samples falling on the borders, one (PbS) gave a positive result and the other (ZnS) gave a negative result.

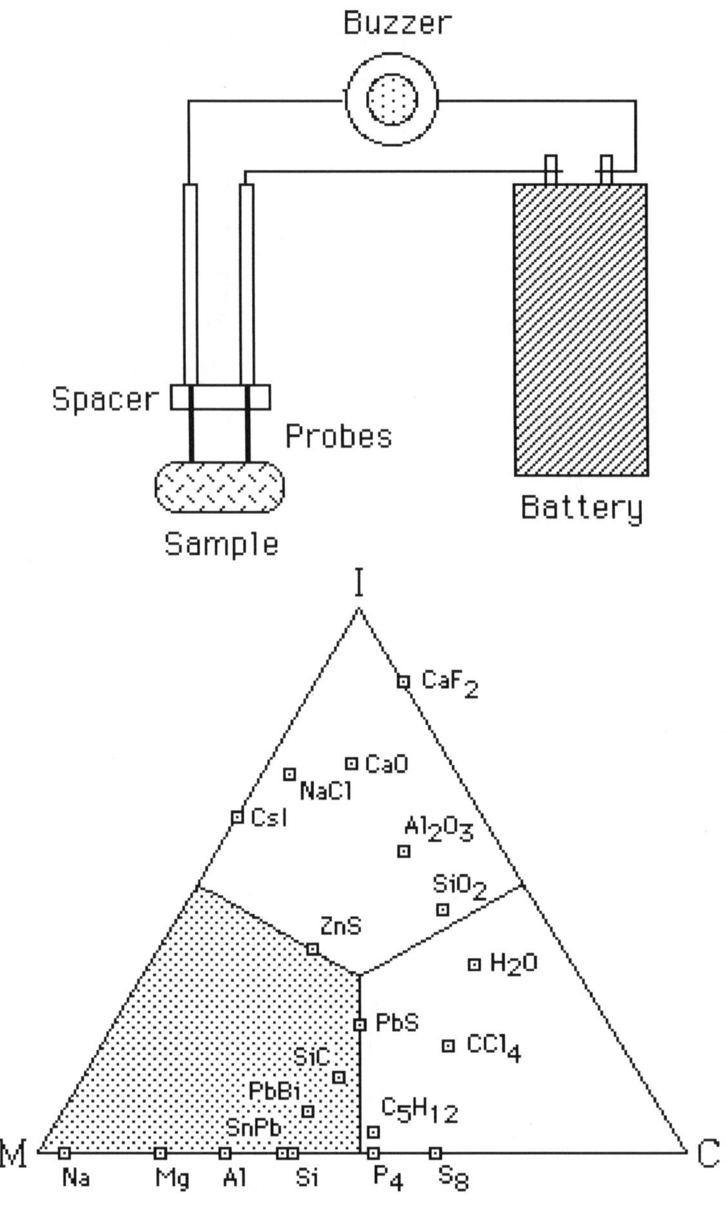

Obviously this demonstration has some serious limitations and should not be pressed too far. In the first place, conductivity is actually not an either/or phenomenon which can be used to unambiguously label all materials as either conductors or insulators. In reality, like bond type, it can take on a continuous range of numerical values. Our either/or result is an artifact of our crude conductivity tester, which requires a certain threshold value for the conductivity before it will produce an audible sound with the buzzer – a threshold value that will change with both the voltage of the battery and the size of the gap between the two probes. In addition, a positive result does not tell us whether we are dealing with a true metallic conductor (negative temperature coefficient for conductivity) or with a very good semiconductor (positive temperature coefficient for conductivity).

Second, our choice of samples is limited by the requirements that they be either liquids or solids at room temperature and pressure (RTP) and, in the latter case, that they also be in the form of large crystals or bulk samples rather than in the form of fine powders. The first of these requirements accounts for the absence of samples in the lower right-hand corner of the covalent extreme, as most simple substances and binary compounds in this region are gases at RTP, whereas the second necessitates that in many cases we have to test large mineral samples (e.g. quartz, sphalerite, fluorite, galena, alumina boules, etc.) rather than standard powdered laboratory reagents.

Third, in order for the demonstration to be useful for the average chemistry teacher, we have to employ samples that are readily available. This requirement accounts for the absence of examples in the upper left-hand area of the metallic extreme, since most of the Zintl phases falling in this region are not found in the typical school stockroom and are not normally available as either macro crystals or bulk samples. In order to compensate for this lack of true intermetallic compounds we have had to employ some common alloys, such as pewter for Sn-Pb bonds and Wood's alloy for Pb-Bi bonds, though these materials are more accurately viewed as solid solutions.

Fourth and lastly, though we have obeyed the constraint against mixing main-block and transition-block samples in the same diagram, we have violated the constraint against mixing samples of varying stoichiometry in order to obtain a sufficiently broad range of examples. Since, according to equation 4.6, bulk conductivity will depend on both bond type and structure type at constant T and P, we would anticipate some anomalies for a correlation based solely on bond-type, since structural variations have not been properly constrained. This may be illustrated in the case of the element C, which, if it were included in the above diagram, would lie to right of P_4 along the bottom edge in the predominantly covalent region. Testing it in the form of the 4/4 infinite-framework structure found in diamond would give a negative conductivity result in keeping with the above sorting map, but testing it in the form of the 3/3 infinite-layer structure found in graphite would give a positive result at variance with the above boundaries. However, in spite of its obvious limitations, it is felt that this demonstration still has some value when used in an elementary course to suggest broad qualitative trends linking bond type, on the one hand, with bulk electrical conductivity, on the other.

Finally it should be noted that when we are dealing with the superposition of a sorting map on the underlying bond triangle – in which case we are concerned with the correlation between certain class properties of specific substances and their underlying bond characters, rather than with just bond type alone – it is appropriate to label the data points using the corresponding formulas of the species that are being sorted.

9.6 References and Notes

1. For additional background on sorting maps, see W. B. Jensen, "Classification, Symmetry and the Periodic Table," in I. Hargattai, Ed., *Symmetry: Unifying Human Understanding*, Pergamon Press: New York, NY, 1986, pp. 487-510.

2. V. M. Goldschmidt, "Drei Vorträge über Geochemie," *Geol. För. Stockh. Förh.*, **1934,** *56*, 385-427 (Diagram on p. 409).

3. L. S. Darken, R. W. Gurry, *The Physical Chemistry of Metals,* McGraw-Hill: New York, NY, 1953, pp. 87-89.

4. W. B. Jensen, *The Lewis Acid-Base Concepts: An Overview*, Wiley-Interscience: New York, NY, 1980, pp. 266-270, 273-274, 280-284.

5. P. P. Fedorov and P. I. Federov, "The Formation of Compounds in Binary Systems

Comprising Gallium (III), Indium (III), and Thallium (I) Chlorides," *Zhur, Neorg. Khim.*, **1974**, *19*, 215-220.

6. E. Mooser, W. B. Pearson, "On the Crystal Chemistry of Normal Valence Compounds," *Acta Cryst.*, **1959**, *12*, 1015-1022. As is apparent from the title, these authors used the term "normal valence compounds" to denote what we have instead called simple homodesmic compounds.

7. W. B. Jensen, "Crystal Coordination Formulas: A Flexible Notation for the Interpretation of Solid-State Structures," in D. G. Pettifor, F. R. de Boer, Eds., *The Structures of Binary Compounds*, Elsevier: Amsterdam, 1989, pp 105-146.

8. N. A. Goryunova, *The Chemistry of Diamond-Like Semiconductors*, MIT Press: Cambridge, MA, 1963, p. 11.

9. J. C. Phillips, *Covalent Bonding in Crystal, Molecules, and Polymers*, University of Chicago Press: Chicago, IL, 1969, p. 150. See also J. C. Phillips, *Bonds and Bands in Semiconductors*, Academic Press: New York, NY, 1973, p. 43.

10. O. Muller, R. Roy, *The Major Ternary Structural Families*, Springer Verlag: Berlin, 1974, pp. 76-77, 140, 142-145, 147, 215-220

11. See, for example, J. K. Burdett, G. D. Price, S. L. Price, "Role of the Crystal-Field Theory in Determining the Structures of Spinels," *J. Am. Chem. Soc.*, **1982**, *104*, 92-95 and the additional references given there in footnotes 7-10, most of which refer to papers published in the solid-state physics literature.

12. W. B. Jensen, "Bond-Type and Structure Sorting Maps," Symposium on Chemical Bonding, Department of Chemistry, University of Wisconsin, Madison, WI, July 1980.

13. W. B. Jensen, "A Quantitative van Arkel Diagram," *J. Chem. Educ.*, **1995**, *72*, 395-398.

14. C. K Jørgensen, *Orbitals in Atoms and Molecules*, Academic Press: London, 1962, p. 84.

15. J. P. Suchet, *Chemical Physics of Semiconductors*, Van Nostrand: New York, NY, 1965.

16. E. Parthé, *Crystal Chemistry of Tetrahedral Structures*, Gordon and Breach: London, 1964.

17. R. M. Rose, L. A. Shepard, J. Wulff, *The Structure and Properties of Materials:*

Vol. IV. Electronic Properties, Wiley: New York, NY, 1966, p. 95.

18. F. R. Bischowski, "The Color of Inorganic Compounds," *J. Am. Chem. Soc.*, **1918**, *40*, 500-508.

19. K. S. Pitzer, J. H. Hildebrand, "Color and Bond Character," *J. Am. Chem. Soc.*, **1941**, *63*, 2472-2475.

20. T. B. Brill, *Light: Its Interaction with Art and Antiquities*, Plenum: New York, NY, 1980, pp. 202-205.

21. K. Nassau, *The Physics and Chemistry of Color*, Wiley-Interscience: New York, NY,1983, Chapters 7 and 8.

22. D. R. Lide, Ed., *CRC Handbook of Chemistry and Physics*, 87rd ed., CRC: Francis and Taylor: Boca Raton, FL, 2007, Section 4, pp. 45-101.

X
Some Possible Alternatives

10.1 Other Choices

In the previous lecture we explored the possibility of constructing sorting maps for various molecular properties based on the general function:

$$Properties = f(FEN_A, FEN_B, b/a, m, T, P) \quad [1]$$

and further demonstrated that it was possible, in the case where the property in question corresponded to molecular structure, to impose the resulting sorting maps directly on the bond-type triangle itself, provided that one kept the variables of stoichiometry (b/a), block type (m), temperature (T) and pressure (P) constant and dealt only with examples of homodesmic binaries and simple substances containing a single type of bond that could be uniquely characterized on the bond-type triangle using the corresponding *FEN* values of its component atoms:

$$Structure = f(FEN_A, FEN_B)_{b/a,m,P,T} \quad [2]$$

This success suggested that it might be of interest to explore other, alternative, structure-sorting maps in which the active variables in equation 1 are different from FEN_A and FEN_B. Of these various possible alternatives, it is the variation corresponding to the active variables FEN_A and b/a which ultimately proved to be of most interest (1):

$$Structure = f(FEN_A, b/a)_{FEN_B,m,P,T} \quad [3]$$

The usual restriction of the sorting set to homodesmic binaries and simple substances, allows one to further apply equation 6 of Section 5.2:

$$b/a = e_A/v_B \qquad [4]$$

and to explicitly substitute the ratio of the valence-electron count (e_A) for atom A to the valence-vacancy count (v_B) of atom B in place of the corresponding ratio of the stoichiometric coefficients (b/a) in function 3:

$$Structure = f(FEN_A, e_A/v_B)_{FEN_B, m, P, T} \qquad [5]$$

Application of function 5 to the homodesmic fluorides of the main-block elements is shown in the following figure, where the requisite values of e_A/v_B and FEN_A have been plotted along the y-axis and x-axis respectively:

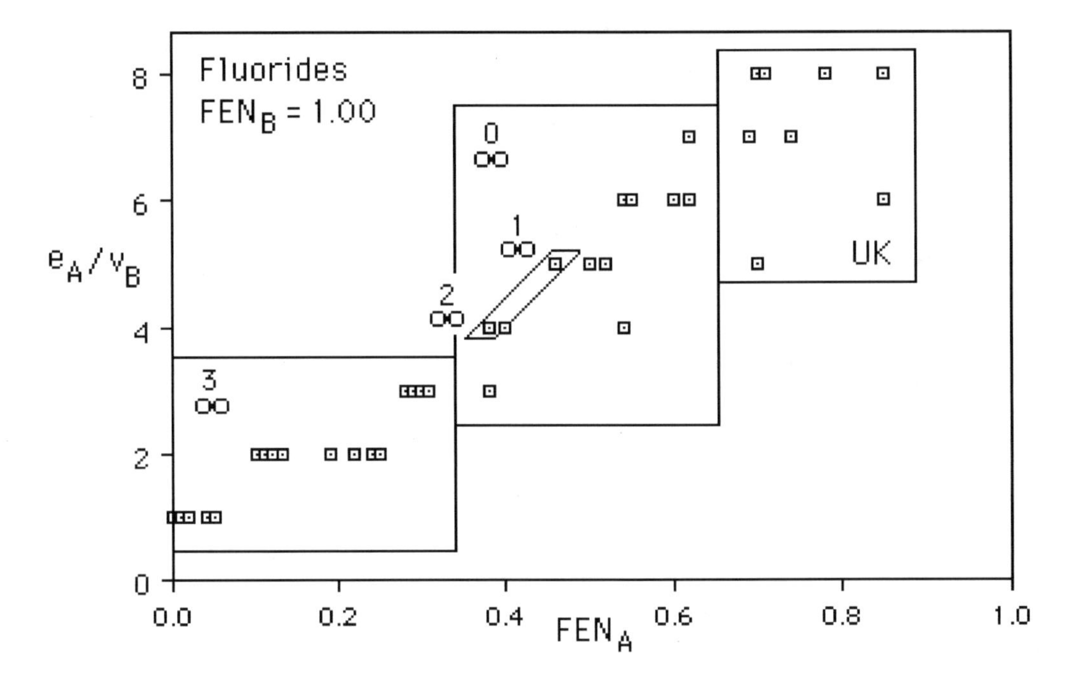

As may be seen, the resulting structure sorting does not correspond to a resolution of alternative structures for a fixed stoichiometry, as was the case of those superimposed on the bond-type triangle, but rather to a much cruder sorting of the structures, as a function of variable stoichiometry, into the categories of infinitely polymerized framework structures (3∞), infinitely polymerized layer structures (2∞), infinitely polymerized chain structures (1∞), and discrete molecular structures (0∞). The first of these categories is localized in the lower left-hand corner and the fourth in the upper right-hand corner, with the infinite chain and layer structures forming a transitional band in the center. In addition, as revealed by the systemic surveys in Lectures 5-8, many potential homodesmic binaries for the strongly electronegative atoms in Groups *5/3 - 8/0* are at present unknown. These have also been sorted into a separate category in the far upper right-hand corner of the diagram and labeled UK for unknown, though their location strongly suggests that, were they ever successfully prepared, they would almost certainly have discrete molecular structures.

The corresponding sorting maps for the main-block homodesmic oxides, chlorides, and hydrides are shown in the following three figures:

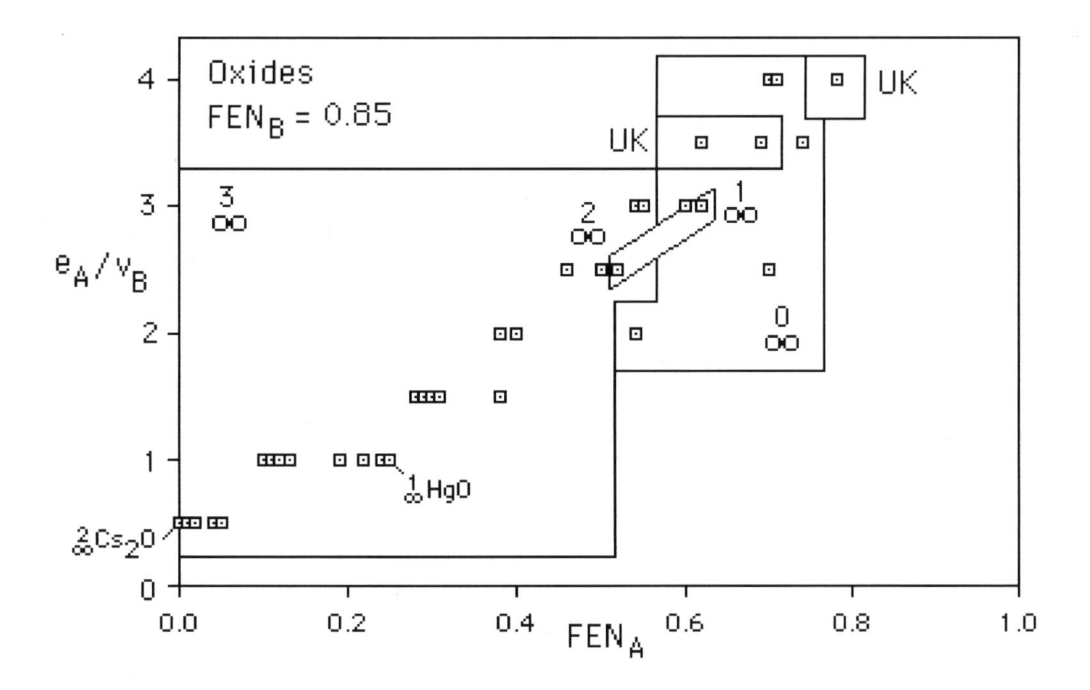

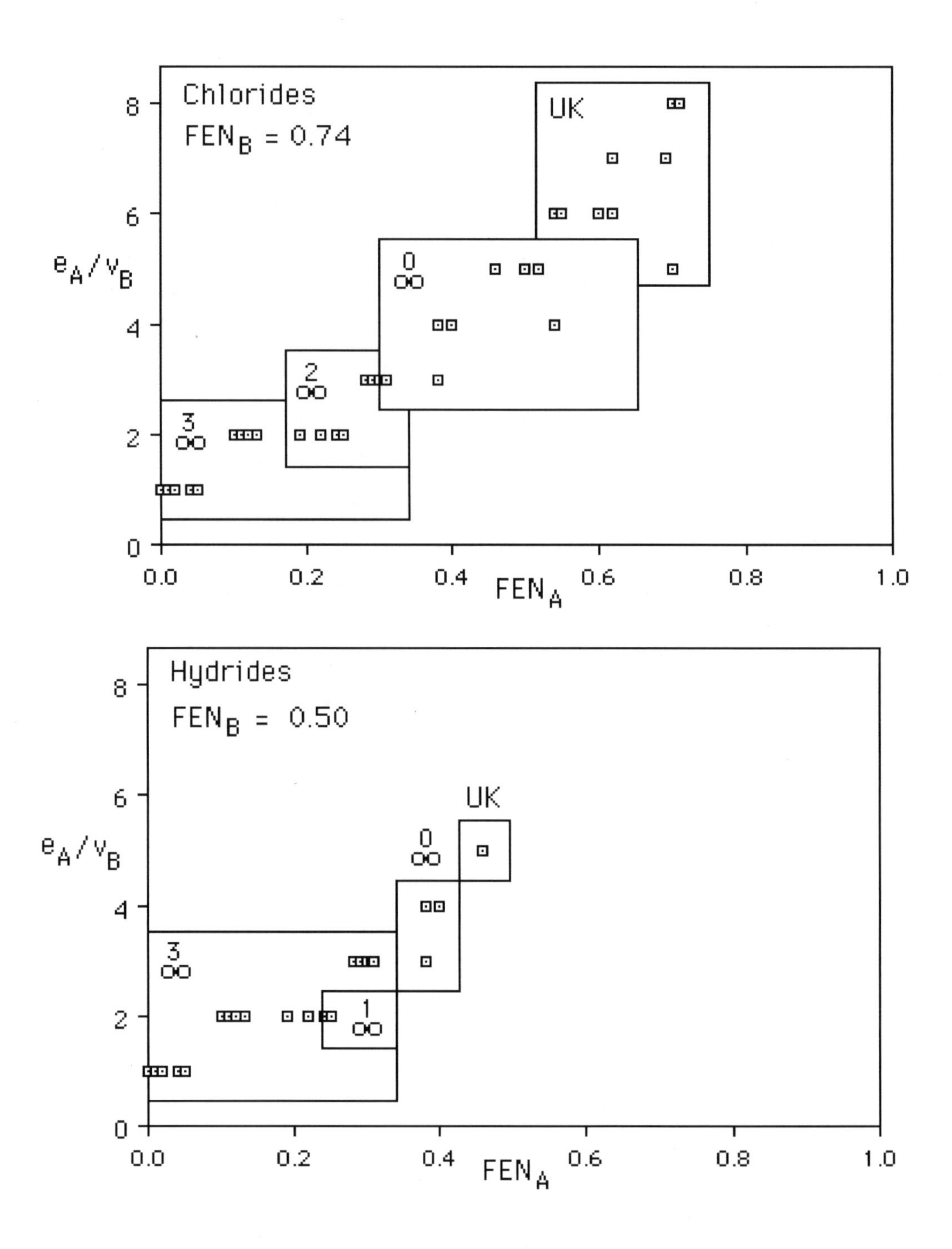

These exhibit the same overall pattern or progression on moving from the lower left-hand corner to the upper right-hand corner as did the sorting map for the fluorides, save that they extend for a shorter and shorter distance along the x-axis. This progressive shortening is a consequence of the requirement that FEN_B must be greater or equal to FEN_A, so as FEN_B decreases in the sequence F > O > Cl > H, both the number of eligible A atoms and the number of potential homodesmic compounds also decrease. Also of note is the fact that the structures of both Cs_2O (infinite layer) and HgO (infinite chain) violate the overall trends for the oxides. Indeed, both structures are atypical of those observed for the other homodesmic oxides in Groups *1/7* and *2/6'* of the periodic table and may be the result of the operation of strong relativistic effects for both elements (2).

10.2 Triangles versus Rectangles

The e_A/v_B - FEN_A sorting maps discussed in the previous section may be visually related to the FEN_A - FEN_B modification of the sorting map of Yeh, employed in the previous lecture, by viewing them both as alternative two-dimensional cross-sections of a three-dimensional right-angle prism based on the coordinates FEN_A, FEN_B, and e_A/v_B:

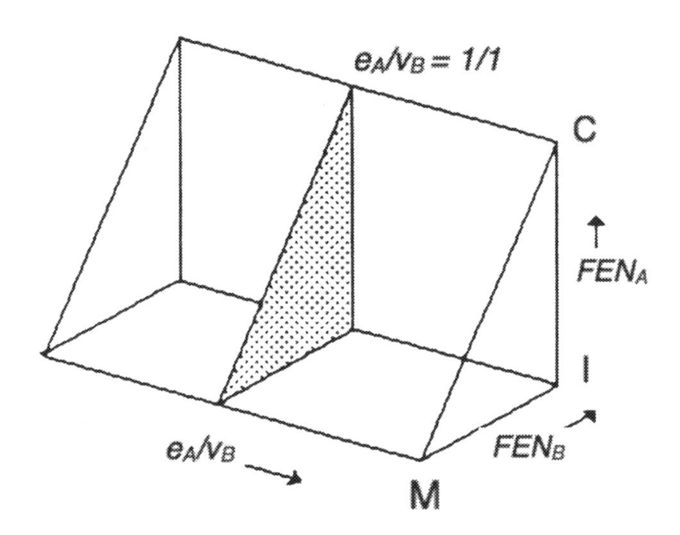

As shown in the above figure, our FEN_A - FEN_B triangular sorting maps, based on the Yeh version of the right-angle, bond-type triangle, may be viewed as a triangular cross-section of this prism located at the point where e_A/v_B = 1/1 or to a stoichiometry of b/a = 1/1. Trigonal sorting maps for other stoichiometries, whose existence was briefly mentioned, but not explicitly developed in the previous lecture, would correspond to similar trigonal cross-sections located at various other key values along the e_A/v_B coordinate.

Likewise, the dimensionality sorting maps given in the previous section may be viewed as various rectangular cross-sections of this prism, each of which cuts our original triangular bond-type triangle along the electronegative axis corresponding to the FEN_B value of the B atom in question:

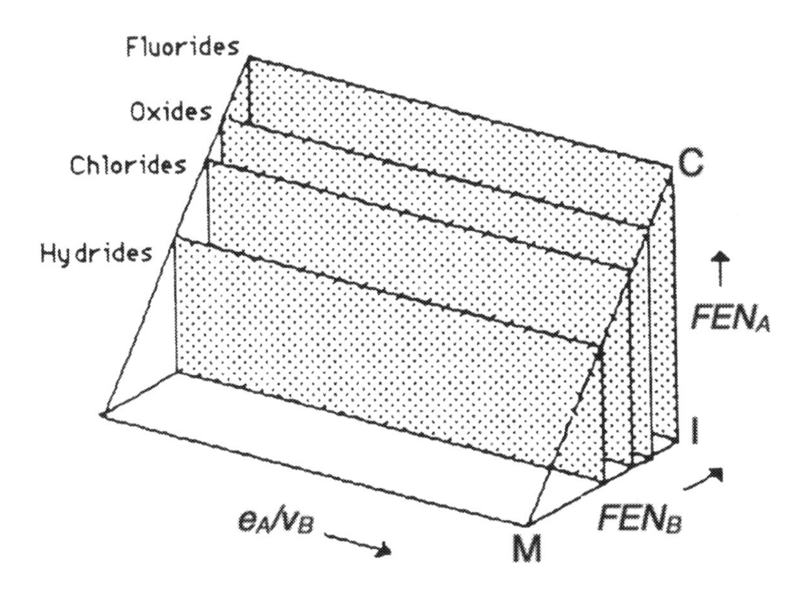

Comparison with the actual plots in Section 10.1 shows that we have rotated each rectangular cross section 90° and have greatly stretched the FEN_A axis in order to more clearly separate the various data points.

These relationships are mathematically similar to those used in constructing ternary phase diagrams, where the corresponding independent variables are mole fraction A (x_A), mole fraction B (x_B), and temperature (T).

In this case it is further assumed that the pressure (P) is kept constant, and that the mole fraction of C (x_C) is dependent on those of A and B. This result can also be represented as a trigonal prism having the coordinates x_A, x_B and T (3). Conventional triangular ternary phase diagrams are then seen to represent various triangular cross-sections of this prism at various fixed values of T, whereas conventional binary composition-temperature phase diagrams represent various rectangular cross-sections of this prism at various fixed values for the mole fraction of the third component.

10.3 Other Uses of *e/v*

The alternative rectangular dimensionality sorting maps presented in Section 10.1 also allows us to more easily visualize the relationship between the bond-type triangle and the conventional periodic table. Each separate electronic block (H-He, main, transition, inner-transition) of the periodic table may be viewed as a rectangular array produced by plotting the inverse of the quantum number for the outer valence shell (*1/n*) versus the *e/v* values of the various elements, as illustrated for the case of the main-block elements in the

1/n ↑									
	Li	Be		B	C	N	O	F	Ne
	Na	Mg		Al	Si	P	S	Cl	Ar
	K	Ca	Zn	Ga	Ge	As	Se	Br	Kr
	Rb	Sr	Cd	In	Sn	Sb	Te	I	Xe
	Cs	Ba	Hg	Tl	Pb	Bi	Po	At	Rn
	Fr	Ra							
	1/7	2/6	2/6'	3/5	4/4	5/3	6/2	7/1	8/0

e/v ⟶

figure at the bottom of the precious page. Yet a second chart of interest can be produced by plotting the *FEN* value of each element versus the *e/v* value of its valence manifold as shown below:

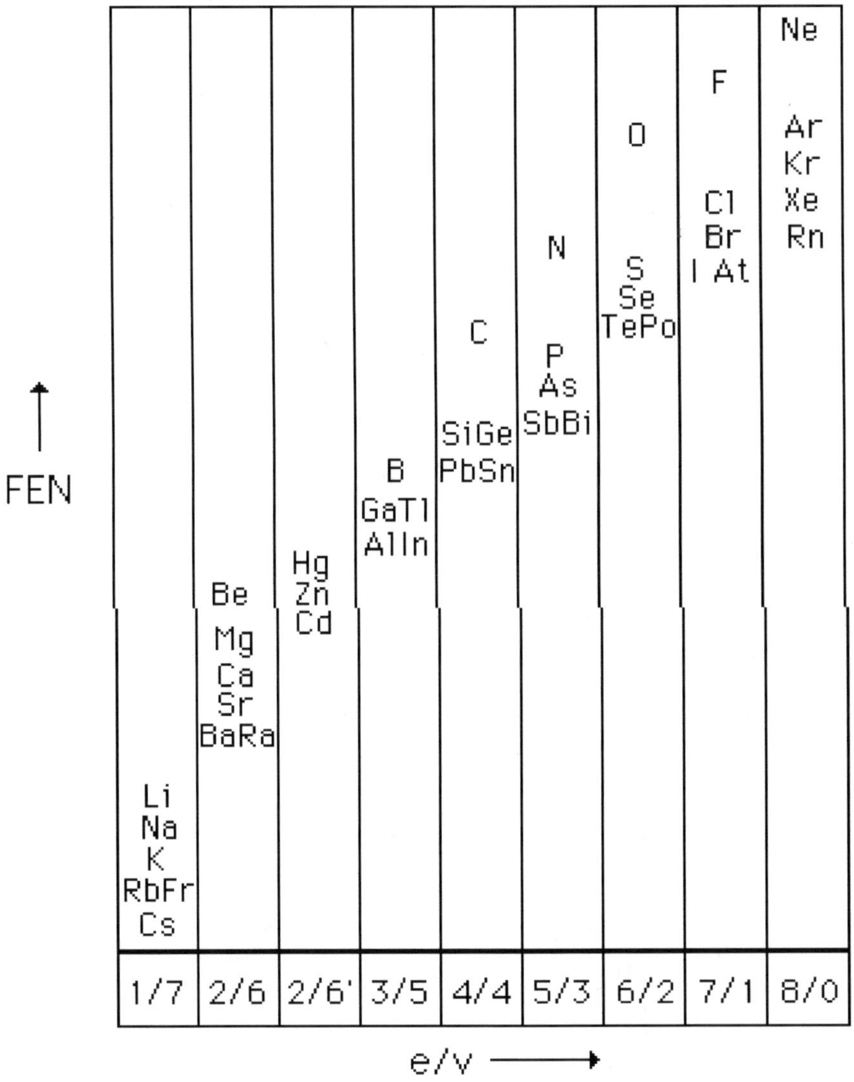

1/7	2/6	2/6'	3/5	4/4	5/3	6/2	7/1	8/0
Li	Be	Hg	B	C	N	O	F	Ne
Na	Mg	Zn	GaTl	SiGe	P	S	Cl	Ar
K	Ca	Cd	AlIn	PbSn	As	Se	Br	Kr
RbFr	Sr				SbBi	TePo	I At	Xe
Cs	BaRa							Rn

FEN ↑

e/v →

10.4 References and Notes

1. Based on W. B. Jensen, "From van Arkel Triangles to Yeh Prisms: Bond-Type Diagrams and Structure Sorting Maps," Inorganic Seminar, Department of Chemistry,

University of Cincinnati, Cincinnati, OH, 14 November 1995.

2. L. J. Norrby, "Why is Mercury Liquid? Or Why Do Relativistic Effects Not Get into Chemistry Textbooks?," *J. Chem. Educ.*, **1991**, *68*, 110-113

3. J. S. Marsh, *Principles of Phase Diagrams*, McGraw-Hill: New York, NY, 1935, Chapter IV.

XI
Possible Extensions

11.1 The Transition-Block and Variable Valence States

In the previous lectures we have restricted the application of the bond triangle to the main-block elements and their compounds. In this lecture we wish to tentatively explore its possible extension to the transition and inner-transition block elements, as well as to the H-He block. These possible applications are still largely exploratory and have not been developed to the same extent as those for the main-block elements.

One of the most salient differences between the main-block versus the transition-block elements is the manner in which they accommodate non-bonding valence electrons. In the case of the main-block elements these electrons remain part of the outer valence shell, where they usually function as stereoactive lone pairs (see however Section 11.4), whereas in the case of the transition-block elements they are stored in the inner (n-1)-shell as stereo-inactive d-electrons. In short, in one case they are stored, so-to-speak, externally, whereas in the other they are stored internally.

As G. N. Lewis observed more than 80 years ago, this means that changes in the oxidation or valence states of these two class of elements have very different origins (1):

Many years ago I was led to the conclusion that when we speak of an element of variable valence we are including two very different types of phenomena, and that such a change as the oxidation of hydrogen sulfide to sulfurous or sulfuric acid is a very different kind of change from the one which occurs when a ferrous salt is converted into a ferric, or a titanous salt into titanic. There are numerous reasons for suspecting a fundamental difference between these two types of oxidation. For example, in the first type oxidation and reduction usually occur in steps of two, while the second type more often involves a change of but one step. Substances which undergo the second type of change are usually colored, while those involved in the first type of change are usually colorless. The distinction between the two types was expressed in my paper of 1916 by referring to the atoms which undergo the second type of valence change as "atoms of "variable kernel."

Thus, in the change from H_2S to SO_2 to SO_3, the number of valence electrons in the outer shell of S does not change, only the number which function as lone pairs versus bonding pairs, whereas in the change from Fe^{2+} to Fe^{3+} there is a fundamental change in the number of electrons found in the inner (n-1)-shell – a difference which Lewis embodied in the idea that, while the atomic cores of the main-block elements remained invariant during chemical change, those of the transition-block were inherently variable.

Unfortunately, in his paper of 1916, and in his subsequent monograph of 1923, Lewis made no attempt to extent his well-known dot symbolism beyond the main-block and, as a consequence, his concept of variable core transition-block elements never made it into the Freshman textbook. However, in 1964 Linnett made a simple proposal for how one might extend traditional Lewis dot diagrams so as to accommodate these elements (2). As in the case of the dot diagrams for the main-block elements, one uses an atom's atomic symbol to represent the invariant noble-gas portion of its atomic core and dots placed around the outer periphery of this symbol to represent the number of electrons in the outer valence shell. However, in the case of the transition elements, one now inserts a circle between these parts in order to suggest an inner (n-1)-shell and places the proper number of dots within this circle to represent the nonbonding (n-1)d electrons stored in the "variable" portion of the atomic core. Thus while Cl can be represented by a single Lewis dot diagram, a separate Lewis diagram is required for each of the six known oxidation or valence states of Mn, as shown below:

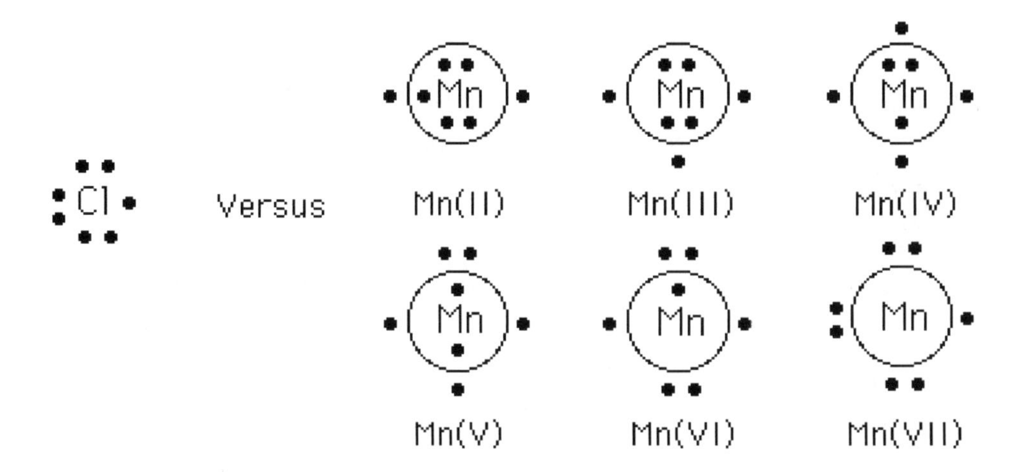

I hasten at this point to remind the reader that, despite the traditional use of the term "Lewis structure," these diagrams in no way purport to represent the actual physical structures of atoms in the same sense as full electron configurations. They are merely a useful symbolic device for counting and partitioning an atom's electrons into chemically significant categories, and for this reason we prefer to use the term "Lewis diagram" instead.

One possible way of extending this symbolism to the inner-transition elements would be to once again reserve the atomic symbol for the innermost noble-gas portion of the atomic core and the outer periphery for the electrons in the active valence manifold, but enclose the variable portion of the core in a double circle in order to suggest that these electrons now reside in inner (n-2)f orbitals rather than in the inner (n-1)d orbitals. Thus we would represent the first three members of the lanthanoids, when in their common M(III) oxidation or valence state, as shown at the top of the next page.

The implications of the above distinctions for how we assign electronegativities to these two classes of elements are profound. As outlined in Appendix I, the Bohr *EN* values used in this monograph are calculated using the

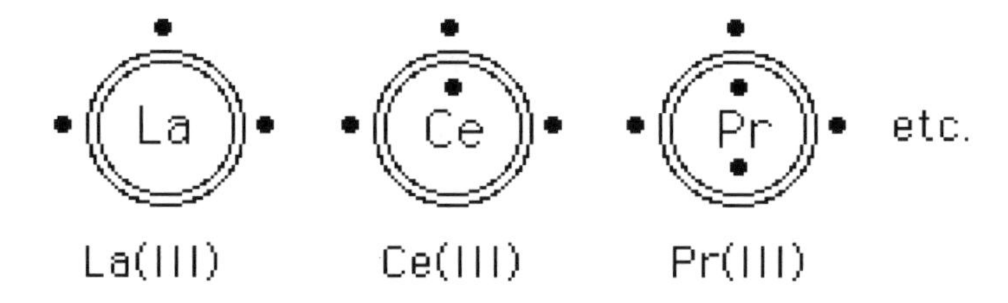

average ionization energies of an atom's valence electrons. Though we have referred to them as "atom electronegativities, they are more accurately characterized as "core" electronegativities, since they obviously measure the average attraction of an atom's core for any surrounding valence electrons. As long as the atom has a single invariant core it is possible, using this definition, to assign it a single invariant electronegativity value. When such an atom forms a molecule or complex ion, there are certainly minor perturbations of this *EN* value due to differences in the electronegativity demands of the various ligands (note that lone pairs can be formally thought of as binding to a hypothetical ligand of zero *EN*) as well as concomitant changes in the orbital hybridization. Though, as discussed in Section 4.6, more sophisticated *EN* definitions are available which can take these perturbations into account, we have found that for purposes of a broad introductory survey of descriptive chemistry, they can, to a first approximation, be ignored.

The case of a transition element is, however, quite different, since here both the atomic cores and the total number of active valence electrons are variable and a separate *EN* value must be calculated for each separate oxidation or valence state. In other words, just as the Mn atom cannot be assigned a single invariant Lewis diagram, so it cannot be assigned a single invariant electronegativity value. As a consequence, tables of *EN* values are given in Appendix I which list multiple *EN* values for each of the transition and inner-transition elements from M(I) up to its maximum possible valence state.

Many aspects of the above discussion were anticipated by C. K. Jørgensen more than 50 years ago (3):

It is worth emphasizing that the electronegativity of a given element increases as a function of "increasing oxidation number" and also to some extent may be a function of the detailed nature of the chemical bonds formed (say carbon in C_2H_6, C_2H_4 and C_2H_2). The former amplification is of minor concern to organic chemistry, but is exceeding important in the transition group compounds. For instance, there is no doubt that Mn(II) has a very low electronegativity, comparable to magnesium, and Mn(VII) has a rather high electronegativity, comparable to gold, and that there does not exist a quantity such as the electronegativity of manganese in general.

If we convert the Bohr electronegativity values for Mn listed in Appendix I into the corresponding *FEN* values using equation 3.1 and compare them with main-block elements having similar *FEN* values, we obtain the results summarized in the following table, where they are further illustrated using the corresponding oxides and oxo complexes (4):

Manganese Oxo Species and their Main-Block Analogs

Species	*FEN*	*Examples*	*Analog*	*FEN*	*Examples*
Mn(II)	0.19	MnO	Mg	0.19	MgO
Mn(III)	0.30	Mn_2O_3	Tl	0.30	Tl_2O_3
Mn(IV)	0.41	MnO_2	Ge	0.40	GeO_2
Mn(V)	0.52	MnO_4^{3-}	P	0.52	PO_4^{3-}
Mn(VI)	0.63	MnO_4^{2-}	S	0.62	SO_4^{2-}
Mn(VII)	0.71	MnO_4^-, Mn_2O_7	Cl	0.74	ClO_4^-, Cl_2O_7

As may be seen, Jørgensen was correct in his comparison of Mn(II) with Mg. However, his comparison of Mn(VII) with Au – which is also a transition metal and hence subject to the same ambiguity of having variable electronegativity values – was a good deal off the mark. In fact, as recognized long ago by Mendeleev, the electronegativity and other chemical properties of Mn(VII) are much closer to those of Cl than to Au.

Not only are the corresponding transition-block and main-block oxo species isovalent, they are also isostructural. Thus MnO and MgO both have a 6/6 NaCl framework structure, Mn_2O_3 and Tl_2O_3 both have a complex C-lanthanoid oxide framework structure, MnO_2 and GeO_2 both have a 6/3 rutile framework structure, the pairs MnO_4^{3-}/PO_4^{3-}, MnO_4^{2-}/SO_4^{2-} and MnO_4^-/ClO_4^- all have discrete tetrahedral structures, and Mn_2O_7 and Cl_2O_7 both have the same discrete bent molecular structure. In addition, both MnO_4^- and ClO_4^- are strong oxidizing agents and both Mn_2O_7 and Cl_2O_7 are highly explosive (4, 5). On the other hand, because of the presence of nonbonding d-electrons and/or empty d-orbitals in their variable cores, the Mn species are all intensely colored and often have magnetic properties quite different from those of their isoelectronic main-block analogs.

As shown below, when we plot the various Mn-O bonds on the Fernelius version of the bond-type triangle, we find, as expected, that they all lie along

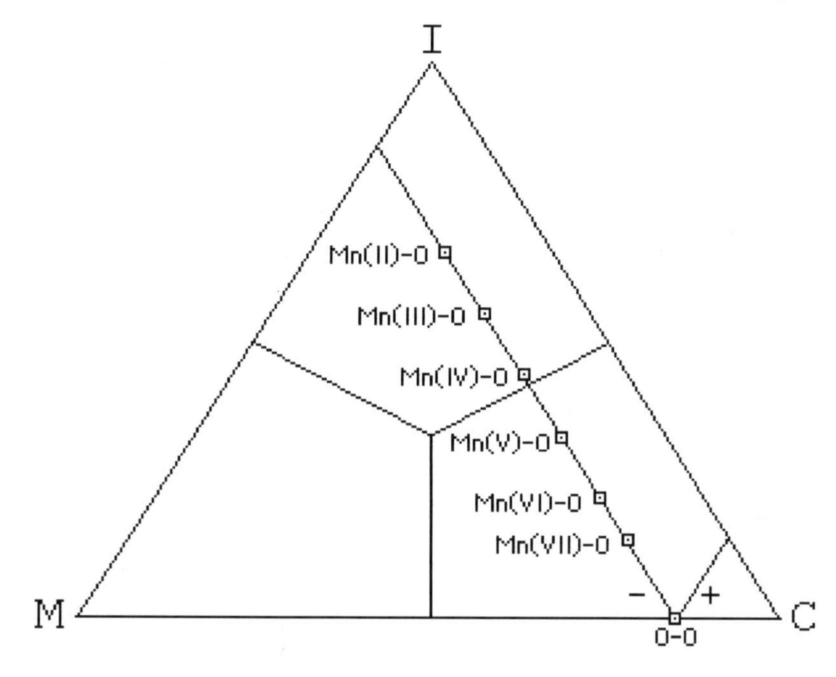

the electronegative branch for oxygen, with those for Mn(II)-O, Mn(III)-O and Mn(IV)-O all lying in the predominantly ionic region, and those for Mn(V)-O, Mn(VI)-O and Mn(VII)-O in the predominantly covalent region. Perhaps no more striking graphical representation of the inherent difference between a typical transition element and a typical main-block element is imaginable. Here, in the variable valence states of a single transition element, we see the same diversity in the properties for the oxide bonds as is displayed by virtually the entire span of the main-block, from Mg-O to Cl-O, and a similar diversity, albeit for smaller spans, is to be expected for the valence states of the other transition elements as well.

11.2 The Transition-Block and the Metallic State

The absence of a single characteristic *FEN* value for Mn means, of course, that the above points for the various Mn-O bonds do not lie on electropositive lines which converge to a single point along the base that uniquely characterizes the Mn-Mn bond in metallic manganese, and the same is equally true for the other transition metals. While the stoichiometry of a homodesmic binary transition metal compound usually allows us to infer the proper oxidation or valence state (and hence the proper *FEN* value) for the transition metal atom, the same is not true in the case of either the simple substances or the case of heterodesmic polycationic binaries which retain a significant degree of M-M bonding (see also the following section).

If we examine the placement along the base of the triangle of the M-M bonds found in main-block metals, we find that they all lie within the predominantly metallic region and extend from the Cs-Cs bond at the metallic vertex, with a *FEN* value of 0 by definition, to the Sn-Sn bond, with a *FEN* value of 0.38, which can exist as a simple substance in either a metallic 6/6 framework structure (white or β-Sn) or in a non-metallic 4/4 framework structure (gray or α-Sn):

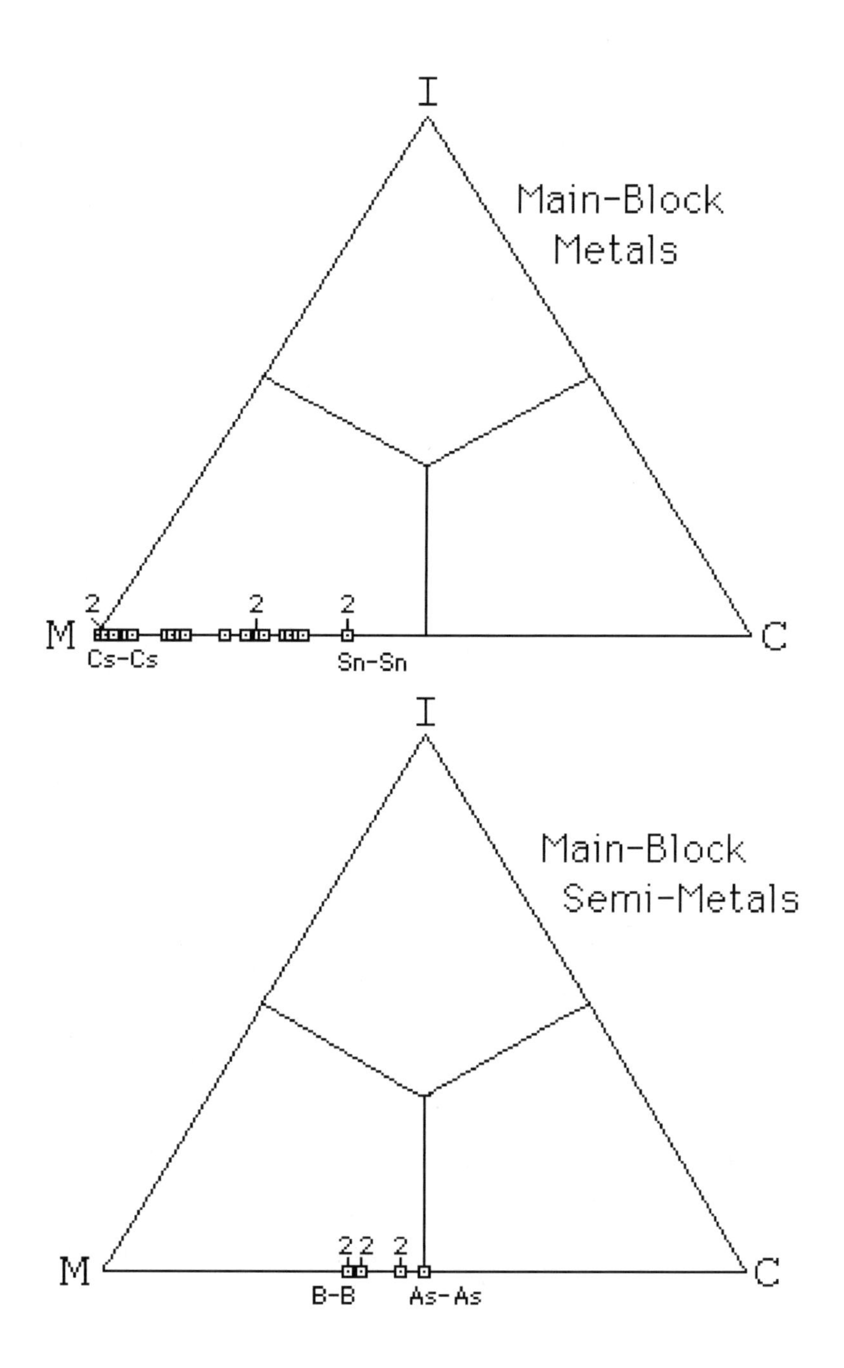

As shown in the second triangle, the bonds for the so-called metalloids or semi-metals also lie just inside the predominantly metallic region and extend from the B-B bond, also with a *FEN* of 0.38, to the As-As bond with a *FEN* of 0.5.

Finally, as might be expected, the nonmetals all lie within the predominantly covalent region and extend from the H-H bond, also with a *FEN* of 0.5, to the F-F bond with a *FEN* of 1.00 by definition:

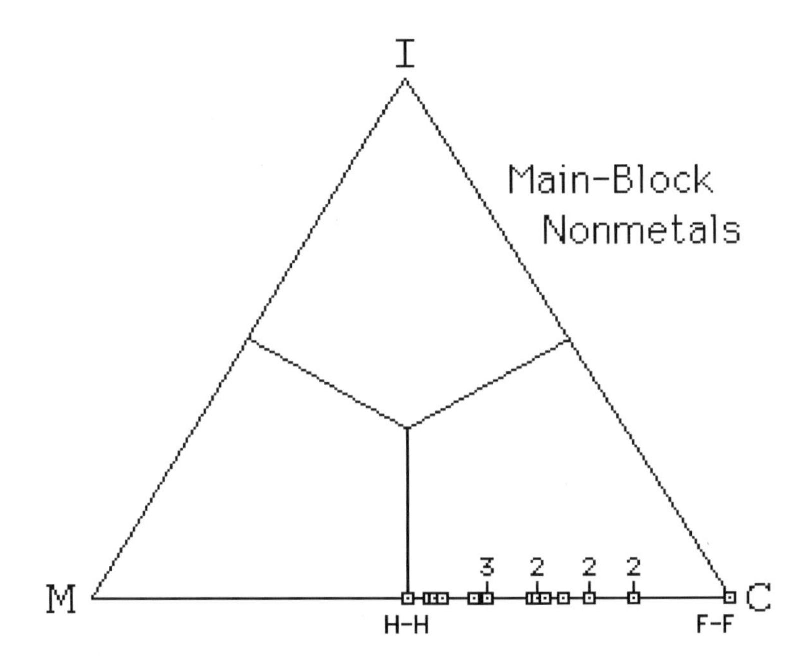

If we convert the *EN* values given for both the transition and inner-transition metals in Appendix I into the corresponding *FEN* values using equation 3.1, we find that, in order for their M-M bonds to lie, in conformity with the above trends, within the predominantly metallic region of the bond diagram, we must assume that the valence state adopted in the bulk transition metals lies between M(II) and M(III), with M(IV) also being a possibility for the inner-transition metals, but for only a few of the heavier transition metals. If we allow the placement of these bonds to also extend into the semi-metallic region, then the upper value can go as high as M(VI) for inner-transition metals and a few of the heavier transition metals. However, empirical experience (albeit limited) using the triangle strongly suggests that the *FEN* values for the M(III) valence state are uniformly the best choice when transition and inner-transition metal-metal bonding is present, unless special evidence indicates otherwise.

While the valence-electron counts and *FEN* values assumed for the main-block elements are in keeping with the simple Lewis model used to rationalize their chemistry at the introductory level, this elementary model, as emphasized in Section 11.1, has never been extensively developed for either the transition or the inner-transition elements. Indeed, detailed treatments of the M-M bonding in the bulk metals for these elements has almost entirely been done within the context of either pseudopotential theory or band theory, and it is not immediately apparent to what extent the results of these treatments are or are not compatible with the simple Lewis picture given here.

11.3 The H-He Block

The only chemically active element in the H-He block is H, which means that its resulting bonds with the elements of the other blocks do not form parallelograms of various sizes, but rather all lie along various sections of the two lines corresponding to the electropositive and electronegative branches which intersect at the location of the H-H bond along the base of the bond-type triangle. Since this bond is located exactly at the midpoint of this base (50% covalent and 50% metallic) these two branches also symmetrically intersect with the remaining two sides of the triangle at their midpoints, which correspond to the H-F bond (50% ionic and 50% covalent) and the Cs-H bond (50% ionic and 50% metallic) respectively.

As may be seen below, all of the bonds in which H is the more electropositive or cationic component, are distributed along the 50% covalency line within the predominantly covalent region of the triangle, with their ionicity values ranging from 50% for the H-F bond to 0% for H-As bond and their metallicity values from 0% for the H-F bond to 50% for the H-As bond. In keeping with these placements, all of the corresponding compounds have discrete molecular structures and are either gases or liquids at RTP:

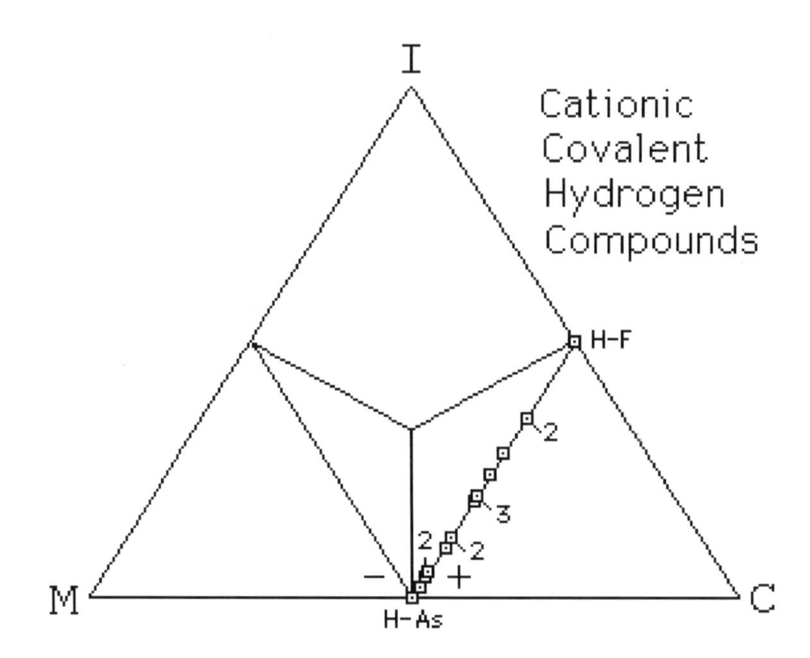

Likewise, bonds in which H is the more electronegative or anionic component are all distributed along the 50% metallicity line and lie within the predominantly metallic region of the triangle. This placement is in keeping with the fact that the bonds to the resulting hydride anion are relatively diffuse, but does not necessarily imply that the resulting compounds as a whole are necessarily metallic. It is customary to further divide the binary anionic hydrides into the three classes of saline, covalent and metallic (6-9). Those of the first class include the hydrides of the Group *1/7* elements or alkali metals and those of the Ca branch (Ca, Sr, Ba, Ra) of the Group *2/6* elements or alkaline-earth metals, with MgH being considered as either an intermediate between this class and the covalent class or as the first member of the latter class.

The location of the resulting bonds for the saline hydrides along the line of the electronegative branch for hydrogen is shown in the following triangle. As may seen, all of them are located in the upper left-hand corner and also lie well above the point for the Mg-H bond, with their ionicity values ranging from 50% by definition for the Cs-H bond to 37% for the Ca-H bond:

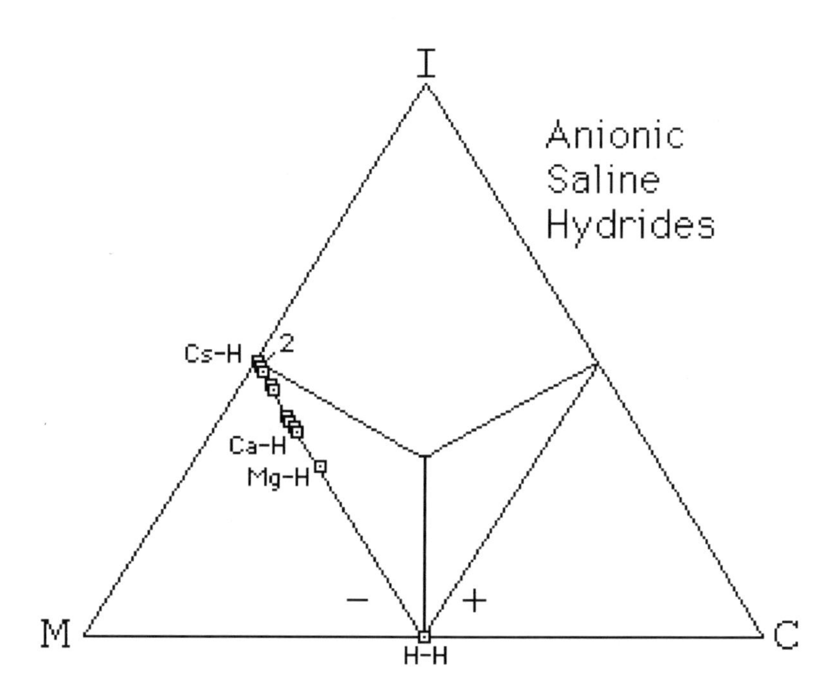

The covalent class of metal hydrides is generally defined so as to consist of BeH and the hydrides of those main-block elements in Groups *3/5* through 5/3 having *FEN* values less than or equal to that of H (*FEN* = 0.5):

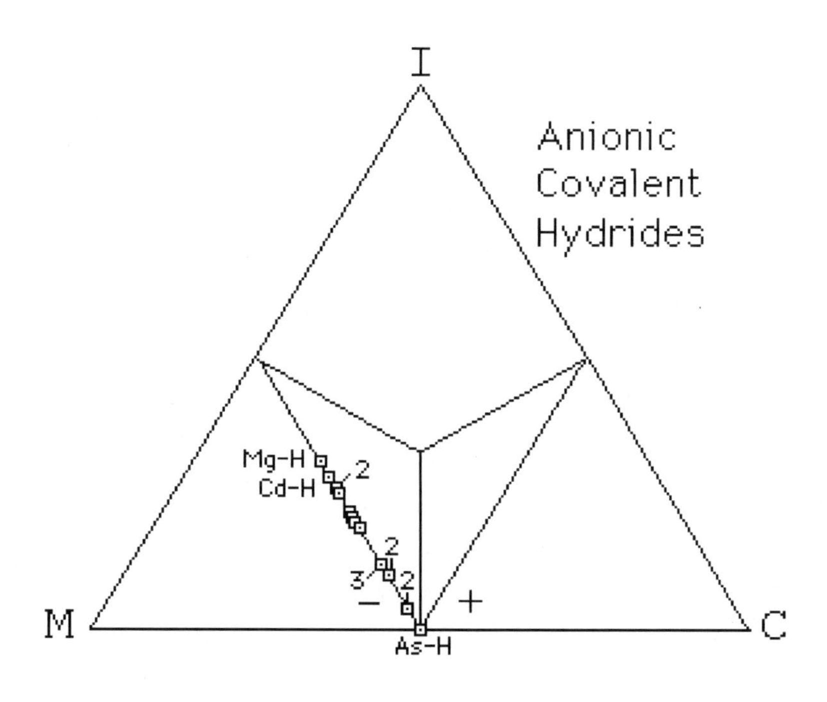

Once again, they are all located along the 50% metallic line, but below, rather than above, the point for the Mg-H bond. Their covalency ranges from 22% for the Cd-H bond to 50% for the As-H bond. Since the *FEN* value for As is identical to that for H, this bond may be considered as either the termination of the electronegative hydride branch (i.e., as As-H) or as the beginning of the electropositive branch (i.e., as H-As)

The metallic hydrides of the third class are all compounds of the transition or inner-transition metals:

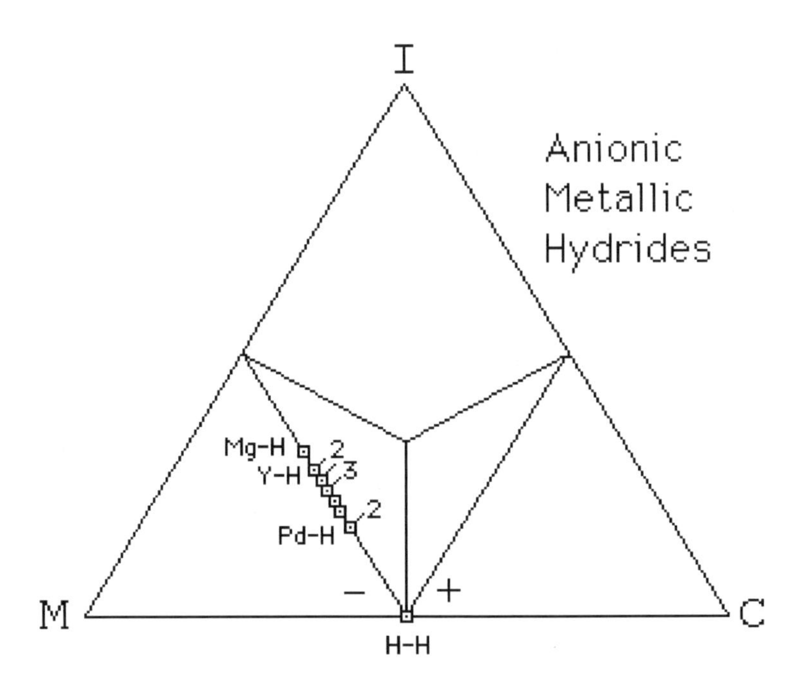

The above triangle (which once again retains the Mg-H point for easy reference) shows the location of the corresponding M-H bonds for those transition metal hydrides that are listed in the monograph by Libowitz (8), and are based on the assumption discussed in the previous section that the corresponding transition-metal cores correspond to the M(III) valence state. As may be seen, all of these bonds overlap with the bonds found among the covalent hydrides and the same would also be true of those for the hydrides of the inner-transition metals. So is this failure to separate these M-H bonds from those of the covalent class a failure of the bond triangle itself? The answer is, of course, no and hearkens back to the necessity, discussed in Section 4.6, of distinguishing between the classification of compounds and the classification of their component bonds. Most metallic transition metal and inner-transition metal hydrides have the idealized composition MH or MH_2 and hence employ only one or two of their three metallic valence electrons in forming the M-H bonds, with the remaining electrons forming highly delocalized M-M bonds. In other words, they are heterodesmic poly-cationic binaries rather than simple homodesmic binaries and the metallic behavior of the compound as a whole is a reflection of the metallic nature of the residual M-M bonding rather than of the relatively covalent nature of the M-H bonds (which, while not destroying the metallic conductivity do, none-theless, alter the mechanical properties of parent metal by inducing brittleness).

Indeed, since the pioneering work of Rundle (10), one of the ways of looking as these compounds is to view them as being formed by the inter-stitial insertion of H atoms into the corresponding close-packed structures of the bulk metals. Here the H atoms extract some electrons from the metal-metal bonds of the bulk metal to form covalent M-H bonds, but not so many as to totally destroy the metal-metal bonding. In keeping with this picture, most of these metallic hydrides exhibit variable composition, with the MH and MH_2 stoichiometries listed above representing idealized limits. The location shown on the bond triangle on the following page of the M-M bonds corresponding to the M-H bonds displayed in the previous triangle, shows that all of them fall within the predominantly metallic region with the metallicity values ranging from a minimum of 67% for the Pd-Pd bonds to a maximum of 78% for the Y-Y bonds.

Finally, it should be noted that, in the case of the inner-transition metals, it is also possible to prepare hydrides having the limiting stoichiometry MH_3 in which, according to to the above model, all of the metal valence electrons have been diverted to the M-H bonds. In keeping with this – and in sharp contrast to the metallic conductivities typical of the MH and MH_2 phases,

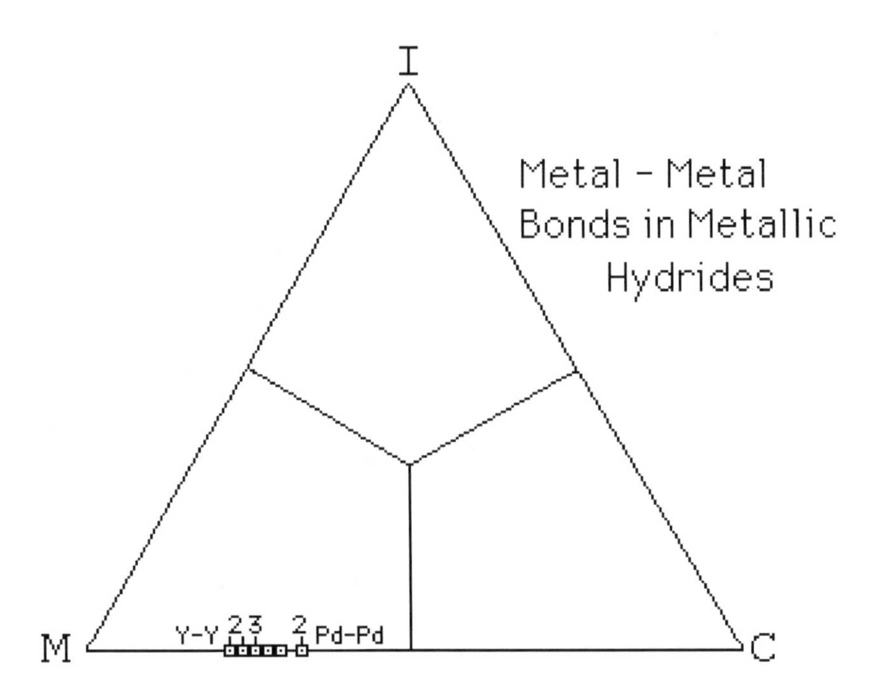

these species are nonmetallic and are usually classified, in accord with the above characterization of their M-H bonds, along with the covalent hydrides (9).

11.4 The Main-Block and Stereoinactive Lone Pairs

With the widespread success in the 1960s of the VSEPR model for the prediction of molecular geometry, attention was focused on the fact that the so-called "inert pair" effect observed for many of the heavier main-block elements often manifested itself not just as a resistance to the formation of higher oxidation states, but also in the occasional formation of stereoinactive lone pairs, which were presumably best described as residing in a spherically symmetrical, unhybridized *ns*-orbital (11). This phenomenon raises the question of whether such stereoinactive lone pairs should be treated the same way as the normal stereoactive lone pairs for these atoms or as forming some type of pseudo-core similar to those found among the transition metals.

In keeping with the latter alternative, the author has actually calculated variable core *EN* values for these elements in Appendix I. On the other hand, in keeping with the first alternative and the fact that our constant *EN* assumption ignores variations in outershell hybridization, does it make sense to treat the case in which the *np*-orbital contribution to that hybridization happens to be zero as somehow exceptional? Thus in Section 9.5, when discussing the placement of the bonds in the bond triangle for the demonstration involving the bulk conductivities of mineral samples, assumption of a constant M(IV) core for Pb resulted in an acceptable placement of the point for the PbS sample on the border of the predominantly metallic region close to the other samples which displayed finite conductivity, whereas assumption of a variable core and use of an *FEN* value for a hypothetical Pb(II) core incorrectly locates the point in the predominantly ionic region above the point for the nonconducting ZnS sample. Similarly, it should be noted that Slater's rules for approximating screening constants make no distinction between *s*- and *p*-electrons when they are located in the same n-shell. In short, the evidence favoring one or the other of these two alternatives is at present ambiguous and a proper resolution will have to await further empirical experience based on the application of these various alternative *EN* values to a broader range of chemical phenomena.

11.5 References and Notes

1. G. N. Lewis, *Valence and the Structure of Atoms and Molecules*, Chemical Catalog Co: New York, NY, 1923, pp. 60-61.

2. J. W. Linnett, *The Electronic Structure of Molecules: A New Approach*, Wiley: New York, NY, 1964, pp. 134-141.

3. C. K. Jørgensen, *Orbitals in Atoms and Molecules*, Academic Press: New York, NY, 1963, p. 80.

4. N. Wiberg, *Inorganic Chemistry*, Academic Press: New York, NY, 2001, pp. 1409, 1412-1415.

5. A. F. Wells, *Structural Inorganic Chemistry*, 4th ed., Oxford University Press: Oxford, 1975.

6. D. T. Hurd, *An Introduction to the Chemistry of the Hydrides*, Wiley: New York, NY, 1952.

7. F. G. A. Stone, *Hydrogen Compounds of the Group IV Elements*, Prentice-Hall: Englewood Cliffs, NJ, 1962.

8. G. G. Libowitz, *The Solid-State Chemistry of Binary Metal Hydrides*, Benjamin: New York, NY, 1965.

9. K. M. Mackay, *Hydrogen Compounds of the Metallic Elements*, Spron: London, 1966.

10. R. E. Rundle, "A New Interpretation of Interstitial Compounds – Metallic Carbides, Nitrides and Oxides of Composition MX," *Acta Cryst.*, **1948**, *1*, 180-187.

11. K. J. Wynne, "Factors Involved in the Stereochemistry of AX_6E Systems of the Heavy Main Group Elements,"*J. Chem. Educ.*, **1973**, *50*, 328-330.

Appendix I
Bohr Electronegativities

If one arranges the chemical elements in an increasing series according to their to their ionization energies, the so-called electropositive elements will be found at the beginning and the electronegative elements at the end.

Johannes Stark, 1903

A1.1 Pauling Electronegativities

Though most present-day chemists are under the mistaken impression that the electronegativity concept was first invented by Linus Pauling in 1932 (1), recent historical studies have in fact shown that the concept predates Pauling by more than a century, having been first introduced into chemistry by Avogadro and Berzelius in the period 1809-1813 (2, 3). Indeed, the relative electronegativity order given by Berzelius for the main-block elements in the 1830s shows a strong linear correlation with that of the present-day Pauling scale, the quality of which (R= 0.97) is comparable to those relating the Pauling scale to other alternative modern definitions of electronegativity.

Berzelius' model of electronegativity, which was based on the theory of imponderable fluids, also postulated a correlation between the electronegativity differences of the atoms forming a compound and the heat released on compound formation, though he himself never developed the full thermochemical implications of this correlation. This correlation was revived once again near the end of the 19th century within the context of the newer mechanical theory of heat and essentially formed the basis of Pauling's attempt to finally quantify the electronegativity scale in the early 1930s.

Pauling's thermochemical scale in based on the correlation equation (4):

$$EN_B - EN_A = k[-\Delta H_f(AB_b)/b]^{0.5} \qquad [1]$$

where k is an empirical constant, $\Delta H_f(AB_b)$ is the heat of formation of some compound AB_b known to contain only A-B single bonds, and one has arbitrarily agreed to assign H an electronegativity value of 2.2. Based on this equation Pauling further proposed the following verbal definition of electronegativity (5):

Electronegativity is the power of an atom within a molecule to attract electrons to itself.

As various critics have repeatedly pointed out over the years, there are at least three fundamental problems with equation 1 and the accompanying verbal definition of electronegativity (6, 7-11):

1. Equation 1 is not a primary definition of electronegativity as verbally defined by Pauling, rather it is, in the words of C. K. Jørgensen, "a secondary effect of something more fundamental" (12). In other words, it is a secondary correlation between electronegativity (as defined by the verbal definition in terms of either a force or potential energy function) and a given molecular property (Π). Indeed, it is but one of many such correlations of the form:

$$\Delta EN = f(\Pi) \qquad [2]$$

not unlike those subsequently found between electronegativity differences and such molecular properties as vibrational force constants, dipole moments, spectroscopic charge-transfer maxima, solid-state work functions, K_{sp} values, polarizabilities, etc. All attempts to derive numerical electronegativities values by working these molecular correlations backwards are subject to precisely the same criticism. In recognition of this, Ferreira has suggested that electronegativity values derived in this fashion should be designated as "secondary electronegativities" in order to distinguish them from those calculated using a primary electronegativity definition which directly measures, in keeping with the verbal definition, either the force or energy of attraction of an atom in a molecule for the surrounding valence electrons (6).

2. Equation 1 lacks a rigorous theoretical basis. Pauling himself originally attempted to justify it by decomposing $\Delta H_f(AB_b)$ into the difference between the energies of atomization of the component pure substances and the resulting binary compound and equating the terms of the resulting thermochemical cycle with the ionic and covalent components of a VB wave function for a polar bond. To assume that there should exist a simple one to one correlation between the components of a macroscopic thermochemical cycle and those of an approximate microscopic wave function was, to say the least, incredibly naive, and subsequent attempts, beginning with that of Mulliken in 1934, to provide a more rigorous justification of the equation, based on the energy terms found in an actual quantum mechanical calculation of typical bond energies, have likewise proven inconclusive, forcing one to agree with Ferreira when he concluded almost 40 years ago that equation 1 essentially had "no quantum mechanical basis" (6).

3. Pauling's verbal definition of electronegativity as a property of atoms within molecules, while certainly in keeping the fact that $\Delta H_f(AB_b$ is a molecular rather than an isolated atomic property, suffers from the problem that the electronic properties of so-called "atoms" within molecules are seldom additive and consequently vary with the number and nature of the other atoms present and with the overall structure of the molecule – a fact which is reflected in the vast electronegativity literature that has since evolved dealing with the manner in which electronegativity values supposedly vary with oxidation state, bonding distance, and the degree of both orbital hybridization and orbital occupancy.

However, despite these problems and the subsequent appearance of more than 25 alternative electronegativity definitions and numerical scales, many of which are not subject to these criticisms, Pauling's original thermochemical scale has remained the scale of choice for most general and inorganic textbooks and has continued to function as the numerical reference "gauge" against which all subsequent scales have been assessed.

A1.2 A New Revised Definition of Electronegativity

In light of the above criticisms, and especially point 3, it would seem logical to abandon Pauling's original verbal definition of electronegativity as a property of atoms within molecules and instead redefine it as a ground-state property of isolated gas-phase atoms:

Electronegativity is a useful average measure of the ability of an isolated atom to attract and retain valence electrons.

This is certainly more in keeping with the manner in which electronegativity is actually used in the chemical literature, as reflected in the fact that, though many of the proposed electronegativity scales which vary with oxidation state, bond length, and degrees of orbital hybridization and occupancy are relatively simple to use, they have so far failed to have a significant impact on the textbook and monograph literature, both of which continue to treat electronegativity as a single-valued parameter unique to each atom type – i.e., as though it is an atomic rather than a molecular property. This atomic interpretation is also inherent in the continued practice of reproducing tables of "intrinsic" electronegativity values and in discussions of the variation of electronegativity across the periodic table as a function of atomic number.

Within the context of this new definition, equations of type 1 and 2 would merely represent interesting empirical correlations between a given property of a molecule and a property of the isolated atoms from which it has been made – empirical correlations that may in fact have only a very approximate underlying theoretical justification and one that most likely is based on a fortuitous cancellation of various complicating factors.

In addition, such a definition would further imply that electronegativity is and will always remain a rather crude semiquantitative concept which, though extremely useful in establishing suggestive, albeit purely empirical, correlations and broad patterns between complex molecular properties and the relative positions of their component atoms within the periodic table, is not itself capable of further theoretical refinement without loosing both its unique conceptual status and its relative usefulness.

A1.3 Electronegativity and Ionization Energies

As indicated in the quote at the beginning of this appendix, it has long been recognized that a simple correlation should between the electronegativities of isolated gas-phase atoms and their corresponding ionization energies. Indeed, some have gone so far as to argue that there is no need for a separate electronegativity scale for gas-phase atoms as the functions of such as scale can be performed just a well by direct reference to the ionization energies themselves (11, 13).

Reasoning that the ability of a monoatomic atom or ion to attract valence electrons should be related in some fashion to the value of its effective nuclear charge, Lehmann and Bähr established an important correlation between ionization energies and electronegativities as early as 1956 using the well-known Bohr equation (14, 15):

$$I = Rhc(Z^*/n)^2 \qquad [3]$$

where I is the ionization energy in units of electron volts, R is Rydberg's constant, h is Planck;s constant, c is the speed of light, Z^* is the effective nuclear charge, and n is the principle quantum number for the electron in question. Solving equation 3 for Z^*/n and evaluating Rhc in units of electron volts, gives:

$$(Z^*/n) = [I/(Rhc)]^{0.5} = 0.27(I)^{0.5} \quad [4]$$

Exploring various choices of I for the main-block elements, and comparing the resulting plots of the period number or n versus Z^*/n with the related plot of n versus Pauling's *EN* values, these authors found that use of the first ionization energy did a poor job of reproducing the observed periodic trends in *EN*, whereas use of the ionization energy for removal of the last valence electron did a much better job. The best fit, however, occurred when they used the function:

$$(Z^*/n)_{av} = 0.27(I_{av})^{0.5} \quad [5]$$

where I_{av} is the average unitless ionization energy (recall the division by *Rhc*) for the valence shell as given by the equation:

$$I_{av} = \Sigma I/e \quad [6]$$

where ΣI is the sum over the successive ionization energies for the entire valence manifold and e, in keeping with our usage in the preceding lectures, is the total number of electrons in the valence manifold, rather than, as per Bohr's original usage, the charge on the electron. Hence the conclusion that:

$$EN = (Z^*/n)_{av} = 0.27(\Sigma I/e)^{0.5} \quad [7]$$

Precisely the same suggestion was made by Martynova and Batsanov in 1980 (16). However, unlike Lehmann and Bähr, who only compared graphs, these latter authors also reported a table of numerical *EN* values calculated in this fashion for both the main-block and transition-block elements.

The use of the average ionization energy in equations 5 and 7, rather than a specific individual ionization value, is keeping with our revised verbal definition, as we are looking for a single-valued numerical characterization of the entire valence manifold of an isolated atom and not just of the behavior of a single valence electron. Consistent with this interpretation, the value of Z^*/n obtained from these equations is obviously an average of some sort for the entire valence manifold, though, unfortunately, the exact nature of that average is not immediately apparent. Consequently, more than two decades ago, the present author tested the related function (17):

$$EN = (Z^*/n)_{av} = 0.27(\Sigma I^{0.5}/e) \qquad [8]$$

in which he used the average of the sum of the square roots of the valence shell ionization energies rather than the square root of the average ionization energy. This function was not only equally effective in reproducing the trends in Pauling's *EN* values for the main-block elements (see below), it also provided a far less ambiguous interpretation of $(Z^*/n)_{av}$ since now:

$$(Z^*/n)_{av} = [\Sigma(Z^*/n)]/e \qquad [9]$$

a result which further simplifies to:

$$[\Sigma(Z^*/n)]/e = (\Sigma Z^*/e)/n \qquad [10]$$

when there is only one valence shell or value of n in the valence manifold.

This definition has a number of important consequences:

1. Though, in keeping with equation 8, the average ability of an atom to attract valence electrons, as measured by the average of the square root of the ionization energy of the valence manifold, is directly proportional to *EN*, electronegativity itself is neither a force nor a potential. Rather it is a dimension-

less constant characteristic of a given atom's overall valence manifold.

2. Since *EN* is, in turn, a function of two dimensionless constants, Z^* and n (recall that Z is not the total nuclear charge but rather the *number* of elementary charges in the nucleus), not only is it easily understood within the context of most introductory chemistry courses, which still make heavy use of simple Bohr theory to interpret atomic structure, it is also relatively easy to rationalize the resulting periodic trends in *EN* in terms of the corresponding periodic trends for Z^* and n. In honor of this fact, the author suggested some years ago that the resulting electronegativity values obtained through application of equation 8 should be called Bohr electronegativities, and he also calculated a complete table of these values for the entire periodic table (17).

A1.4 Trends in the Main Block

A periodic table of Bohr electronegativities for the main-block elements, corresponding to their maximum potential valence values, is shown at the top of the following page, and the manner in which these values correlate with the standard Pauling values in the accompanying graph. As can be seen, the Bohr (EN_B) and Pauling (EN_P) scales are related by the correlation equation:

$$EN_P = 1.564EN_B - 0.043 \qquad [11]$$

This correlation is comparable in quality (R = 0.97) with that which exists, for example, between the Pauling scale and the well-known Allred-Rochow scale (R = 0.96), and can be further approximated without loss of accuracy by the even simpler equation:

$$EN_P \approx 1.5EN_B \qquad [12]$$

	1/7	2/6	2/6'	3/5	4/4	5/3	6/2	7/1	8/0
2	Li 0.63	Be 0.99		B 1.26	C 1.57	N 1.88	O 2.17	F 2.47	Ne 2.79
3	Na 0.61	Mg 0.90		Al 1.09	Si 1.31	P 1.53	S 1.74	Cl 1.96	Ar 2.18
4	K 0.56	Ca 0.79	Zn 0.99	Ga 1.13	Ge 1.31	As 1.50	Se 1.70	Br 1.87	Kr 2.05
5	Rb 0.55	Sr 0.77	Cd 0.96	In 1.08	Sn 1.26	Sb 1.43	Te 1.59	I 1.74	Xe 1.91
6	Cs 0.53	Ba 0.73	Hg 1.02	Tl 1.12	Pb 1.27	Bi 1.42	Po 1.58	At 1.73	Rn 1.89
7	Fr 0.53	Ra 0.73	112	113	114	115	116	117	118

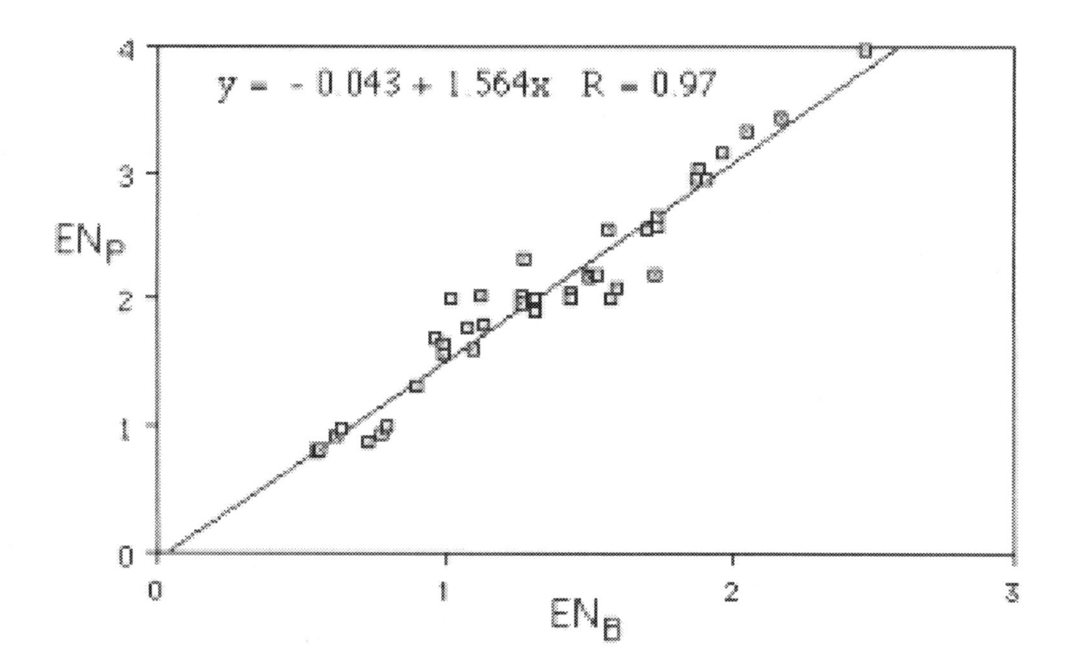

However, since the range of the Pauling scale is purely arbitrary, having been determined by Pauling's decision to set the electronegativity of H at 2.2, there is no compelling reason to shift the values of the Bohr scale to accord with it, and we have instead given the absolute Bohr values calculated by means of equation 8 instead.

Inspection of the above table shows that the values of the Bohr electronegativities for the main-block elements vary from a maximum of 2.79 for Ne in the upper right-hand corner to a minimum of 0.53 for Fr in the lower left-hand corner. Unlike the individual ionization energies, which display marked maxima and minima on passing across a given period of the periodic table, the *EN* values increase uniformly from left to right across a given period, though, in keeping with the effects of both the transition-block and inner-transition block insertions, they also display a significant jump on passing from Group *2/6* to Group *2/6'* in rows 4-6. Similarly, the values decrease on passing down a given group, with the exception of two discontinuities in Groups *2/6'* through *4/4*, which occur on passing from rows 3 to 4 and from rows 5 to 6, again reflecting the effects of both the transition-block and inner-transition block insertions. As also expected, these discontinuities decrease in magnitude on passing from Group *2/6'* to Group *4/4*. These variations are precisely those which we would anticipate from the manner in which the atomic parameters of $(Z^*)_{av}$ and n vary across the main-block.

A1.5 Trends in the Transition Block

As was seen in the previous section, for the main-block elements the Pauling electronegativities correspond to the maximum oxidation-state Bohr electronegativities, and empirical experience with electronegativity correlations shows that the identical electronegativities work for the lower oxidation states of the main-block elements as well. In other words, the identical gas-phase atomic electronegativity value appears to work independently of whether the atom's valence electrons are eventually used for bond formation or are employed as stereoactive lone pairs. However, those main-block elements which display the "inert pair" effect appear to be an exception to this rule. In these cases the nonbonding s-electrons are stereo-inactive and appear to act as "pseudo-core" electrons, and here empirical experience shows that a Bohr electronegativity value based on *e – 2* valence electrons works best. These alternative inert pair values are given below for the elements In, Sn, Tl and Pb, for which this effect is most common:

In(I) = 0.65, Tl(I) = 0.67, Sn(II) = 0.86, Pb(II) = 0.89 [13]

This observation is also pertinent to an interpretation of transition-block electronegativities. Here again, those valence electrons not involved in bond formation are stored as nonstereoactive pseudo-core electrons and it appears necessary to report Bohr electronegativity values for all possible positive oxidation states up the theoretical maximum allowed by the valence manifold criterion for block formation, as has been done in the following three tables. Comparison with Pauling's reported single-valued thermochemical electronegativities for the transition-block elements shows that they correlate best

Σe	Sc	Ti	V	Cr	Mn	Fe	Co	Ni	Cu
1	0.64	0.70	0.70	0.70	0.74	0.76	0.76	0.75	0.75
2	0.83	0.85	0.87	0.90	0.90	0.92	0.94	0.95	0.98
3	1.00	1.04	1.07	1.11	1.12	1.11	1.15	1.17	1.20
4		1.22	1.27	1.30	1.33	1.34	1.35	1.38	1.42
5			1.45	1.50	1.54	1.55	1.57	1.58	1.63
6				1.68	1.73	1.75	1.78	1.80	1.83
7					1.91	1.94	1.98	2.00	2.04
8						2.11	2.17	2.20	2.24
9							2.36	2.39	2.43
10								2.57	2.62
11									[2.81]

Σe	Y	Zr	Nb	Mo	Tc	Ru	Rh	Pd	Ag
1	0.68	0.70	0.71	0.72	0.73	0.73	0.74	0.78	0.74
2	0.81	0.84	0.86	0.90	0.89	0.92	0.94	0.99	1.00
3	0.95	0.99	1.03	1.07	1.08	1.09	1.13	1.17	1.20
4		1.14	1.19	1.26	1.26	1.28	1.30	1.35	1.38
5			1.33	1.43	1.42	1.45	1.49	1.52	1.56
6				1.56	1.58	1.62	1.65	1.69	1.72
7					1.72	1.77	1.81	1.86	1.89
8						1.92	1.96	2.01	2.05
9							2.11	2.16	2.21
10								2.31	2.36
11									[2.54]

Σe	Lu	Hf	Ta	W	Re	Os	Ir	Pt	Au
1	0.67	0.71	0.76	0.76	0.76	0.80	0.82	0.81	0.82
2	0.82	0.88	0.92	0.95	0.93	0.95	0.96	0.99	1.02
3	0.96	1.00	1.03	1.08	1.08	1.09	1.11	1.14	1.17
4		1.12	1.17	1.21	1.22	1.24	1.25	1.29	1.33
5			1.29	1.34	1.37	1.39	1.41	1.43	1.47
6				1.47	1.50	1.53	1.56	1.58	1.61
7					1.63	1.66	1.69	1.73	1.76
8						1.79	1.83	1.86	1.90
9							1.96	2.00	2.04
10								2.12	2.17
11									[2.32]

with the Bohr values for either oxidation state II or III, in keeping with the stoichiometries of the various compounds originally used to calculate the Pauling values.

For some of these elements, the necessary ionization energies for calculating the Bohr values for the highest potential oxidation states are missing and in these cases the reported values were obtained by extrapolation of plots of EN_{max} versus e_{max} across a given series of a given block of the periodic table, as it was discovered that these were strictly linear for all blocks in the table, as shown below for series 1 of the transition-block, which was used to obtain an extrapolated EN_B for Cu(XI) of 2.81:

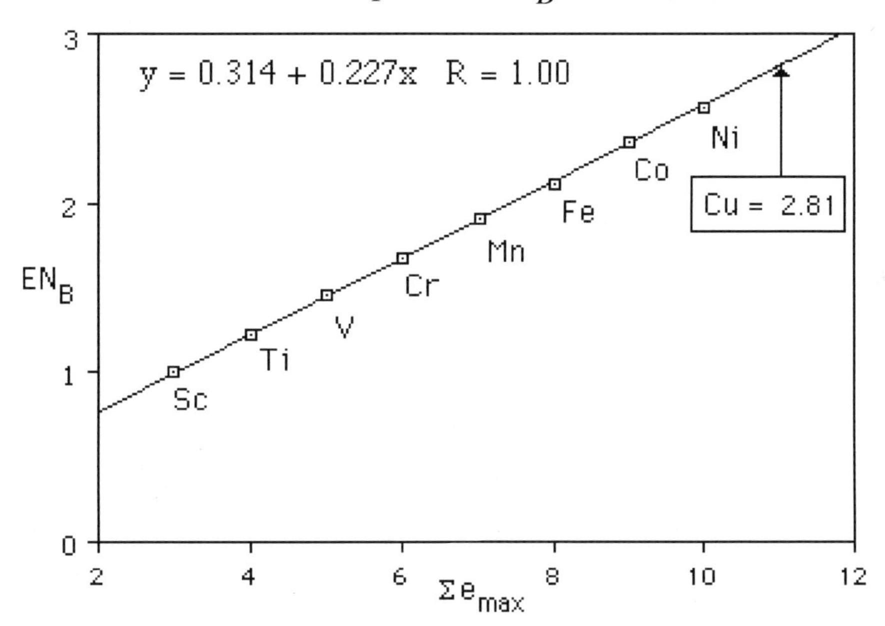

A1.6 Trends in the Inner-Transition Block

In the case of the inner-transition block elements, valence electrons not involved in bond formation are again stored as inner, nonstereoactive, pseudo-core electrons, so we have again reported Bohr electronegativity values in the following table for all possible positive oxidation states up to the maximum allowed by the valence manifold criterion for block formation:

Σe	La	Ce	Pr	Nd	Pm	Sm	Eu	Gd	Tb	Dy	Ho	Er	Tm	Yb
1	0.64	0.63	0.63	0.63	0.64	0.64	0.64	0.67	0.66	0.66	0.66	0.67	0.67	0.68
2	0.77	0.76	0.75	0.76	0.76	0.77	0.77	0.80	0.78	0.79	0.80	0.80	0.80	0.81
3	0.91	0.91	0.92	0.93	0.93	0.95	0.96	0.95	0.94	0.96	0.96	0.96	0.97	0.99
4		1.09	1.11	1.13	1.13	1.14	1.16	1.16	1.13	1.15	1.16	1.16	1.17	1.19
5			1.26	1.28	1.27	1.30	1.32	1.30	1.28	1.30	1.31	1.31	1.32	1.35
6				1.45	1.44	1.46	1.50	1.46	1.44	1.46	1.48	1.47	1.49	1.52
7					1.60	1.63	1.67	1.62	1.60	1.63	1.64	1.63	1.65	1.69
8						1.80	1.85	1.79	1.81	1.79	1.81	1.79	1.82	1.86
9							2.02	1.95	1.92	1.96	1.97	1.96	1.99	2.03
10								2.11	2.08	2.12	2.14	2.12	2.16	2.20
11									2.24	2.28	2.31	2.28	2.32	2.37
12										2.45	2.47	2.45	2.49	2.54
13											2.64	2.61	2.65	2.71
14												2.77	2.82	2.88
15													2.95	3.06
16														3.23

Not surprisingly, the reported Pauling values for these elements correlate best with the Bohr values for oxidation state III. In the case of the lanthanoids, the necessary ionization energies are unknown beyond the value for the removal of the fifth valence electron and so a larger percentage of the reported values have been obtained from extrapolation of the corresponding EN_{max}- e_{max} plots. In the case of the actinoid series, there is even less ionization data available so that use of EN_{max}- e_{max} extrapolations allow us to estimate only an average EN_B for each potential oxidation state and this single value has been assumed to be approximately valid for all of the elements in question:

1	2	3	4	5	6	7	8	9	10	11	12	13	14	15	16
0.67	0.80	0.93	1.07	1.20	1.33	1.47	1.59	1.73	1.86	1.99	2.12	2.25	2.39	2.52	2.65

A1.7 A Problem in the H-He Block

Strict application of equation 8 to H and He yields Bohr electronegativity values of 1.00 and 1.67, respectively. Since H is widely recognized to have an *EN* value slightly less than that for carbon (EN_B = 1.57) and greater than that for boron (EN_B = 1.26), a value of 1.00 appears to be too low to account for its observed chemical behavior in the hydrocarbons versus the boron hydrides. Similarly, the reported value for He also appears to be too low to account for its observed chemical inertness, as other elements still resistant to compound formation under ambient conditions (Ne and Ar) or to compound formation in their highest potential maximum oxidation states (e.g. O and many of the transition and inner-transition elements) all have Bohr electronegativity values grater than 2.1. Consequently the Bohr electronegativity values for both H and He have been multiplied by an empirical factor of 1.5 so as to give the following adjusted values:

H = 1.50, He = 2.51 [14]

The reason for these discrepancies is unknown, but a similar empirical adjustment is required for H in roughly 90% of the electronegativity scales that have been previously proposed.

A1.8 Relationship to the Allen Spectroscopic Definition

Several other empirical gas-phase atomic electronegativity scales based on the use of ionization energies have also been proposed in the past, the most recent of which is due to Allen. In 1989 he proposed a new method of calculating electronegativity values for the main-block elements based on averaging the valence-shell orbital energies of an atom as measured, via Koopman's theorem, by the corresponding one-electron orbital ionization energies (18):

$$EN_S = (mE_p + nE_s)/(m + n) = (mI_p + nI_s)/(m + n) \qquad [15]$$

where m and n are the number of p and s valence electrons, E_p and E_s are their respective one-electron orbital energies, and I_p and I_s are the corresponding one-electron orbital ionization energies. Allen called his new electronegativity values "spectroscopic electronegativities," though he has since changed the name several times. In the case of the main-block elements they display a very high correlation with both the traditional Pauling (R = 0.98) and Allred-Rochow (R = 0.98) electronegativity scales.

It is important to recognize that not only are functions 8 and 15 mathematically distinct, the nature of the ionization averages which appear in them are also quite different, as may be illustrated through their application to an Al atom. In the case of equation 8, we are averaging the square roots of the ionization energies required to successively ionize one valence electron after another to create the Al^{+}, Al^{2+} and Al^{3+} ions, respectively:

$$EN_B = (0.27/eV^{0.5})[(5.985eV)^{0.5} + (18.823eV)^{0.5} + (28.441eV)^{0.5}]/3 = 1.09 \qquad [16]$$

These are, of course, the successive ionization energies commonly listed in typical inorganic textbooks.

In the case of equation 15, however, we are averaging the ionization energies required to create a series of Al^{+} ions, only one of which corresponds to the ground-state ion:

$$EN_S = (2I_s + I_p)/3 = [2(11.315\text{eV}) + 1(5.987\text{eV})]/3 = 9.539\text{eV} \qquad [17]$$

where I_p is the ionization energy required to create a ground state $[Ne]2s^2$ Al^+ ion and is numerically equal to I_1 (5.987eV) in equation 15. I_s, however, is the ionization energy (11.315eV) required to create an excited-state [Ne] $2s^12p^1$ Al^+ ion and is equal to I_1 plus the energy required to excite one of the 2s electrons in the ground-state Al^+ ion to a 2p orbital. Quite obviously, application of equation 15 requires a knowledge not only of ionization energies but of spectroscopic transition energies, which is, of course, why Allen originally referred to his values as "spectroscopic" electronegativities. While the spectroscopic scale has the advantages of being theoretically consistent with MO theory and of being one of the few electronegativity definitions not requiring an empirical adjustment of the value for H, the present author is dubious about both trying to explain its conceptual basis to a typical general chemistry class and of being able provide said students with a simple reductionistic rationale for its observed periodic trends.

In contrast to the *EN* values calculated by means of equation 8, which are dimensionless, those calculated by means of equation 15 have the units

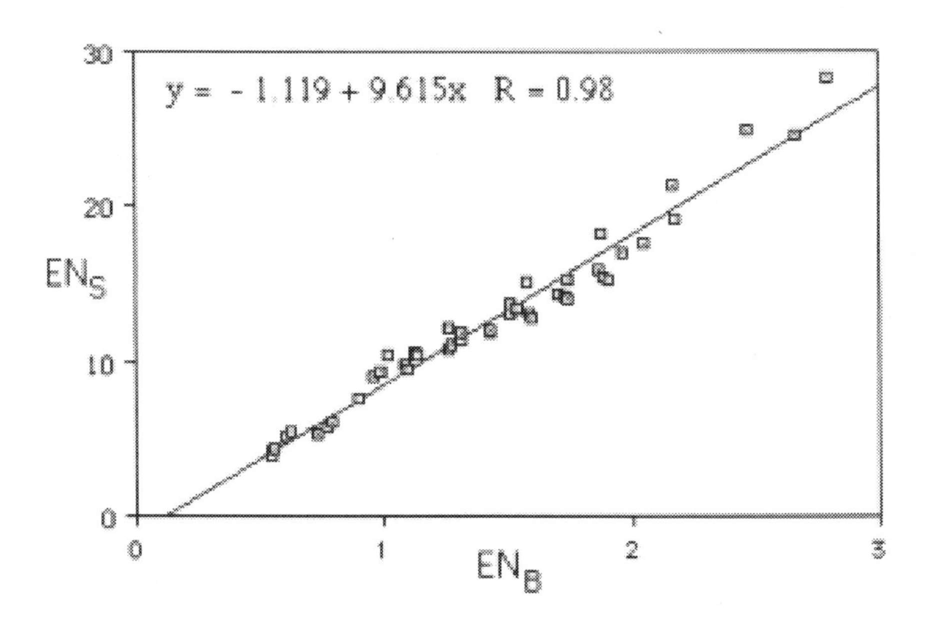

of energy. Nevertheless, as might be anticipated from the fact that both scales display a high degree of linear correlation with the Pauling scale, both scales also display a high degree of linear correlation with one another (R = 0.98) as may be seen from the graph on the previous page. Since publishing his original definition, Allen has also extended it to include the transition-block elements as well and has published tables of numerical EN_S values for both blocks (19).

A1.9 Summary and Conclusions

It has been argued that electronegativity is best understood as a property of the isolated gas-phase atom rather than of the "atom within the molecule," and that such electronegativities can function not only as useful semi-quantitative measures of the average attraction of an atom for its valence electrons but also as useful numerical parameters for correlating trends in various molecular properties with those of their component atoms. In keeping with these assumptions, this appendix has presented an atomic electro-negativity scale based on the well-known Bohr relationship between Z^*/n and ionization energy. The resulting dimensionless Bohr electronegativity values not only display a high degree of correlation with the conventional Pauling values, as well as with other modern electronegativity scales, they are also easy to rationalize at an elementary level in terms of simple Bohr theory and in terms of the corresponding periodic trends for the fundamental parameters Z^* and n.

A1.10 References and Notes

1. L. Pauling, "The Nature of the Chemical Bond: IV. The Energies of Single Bonds and the Relative Electronegativity of Atoms," *J. Amer. Chem. Soc.*, **1932**, *54*, 3570-3582.

2. W. B. Jensen, "Electronegativity from Avogadro to Pauling: I. Origins of the

Electronegativity Concept," *J. Chem. Educ.*, **1996**, *73*, 11-20.

3. W. B. Jensen, "Electronegativity from Avogadro to Pauling: II. Late 19th and Early 20th Century Developments," *J. Chem. Educ.*, **2000**, *80*, 279-287.

4. Pauling originally used single-bond energies derived from the dissociation of gaseous diatomic molecules. However, in this form the equation could only be applied to a few elements and its extension to the rest of the periodic table required the use of average heats of formation per bond instead, as defined by the version of the correlation equation given here.

5. L. Pauling, *Nature of the Chemical Bond*, Cornell University Press: Ithaca, NY, 1939, p. 58.

6. R. Ferreira, "Electronegativity and Chemical Bonding," *Adv. Chem. Phys.*, **1967**, *13*, 55-84.

7. V. P. Spiridonov, M. Tatevskii, "The Concept of Electronegativity. I. An Analysis of the Definitions of Electronegativity Given by Various Authors," *Russ. J. Phys. Chem.*, **1963**, *37*, 522-526.

8. V. P. Spiridonov, M. Tatevskii, "The Concept of Electronegativity. II. An Analysis of Pauliing's Scale of Electronegativities," *Russ. J. Phys. Chem.*, **1963**, *37*, 661-664.

9. V. P. Spiridonov, M. Tatevskii, ""The Concept of Electronegativity. III. Examination of Different Empirical Methods of Determining Numerical Values of Electronegativity," *Russ. J. Phys. Chem.*, **1963**, *37*, 848-849.

10., V. P. Spiridonov, M. Tatevskii, "The Concept of Electronegativity. IV. Theoretical and Semiempirical Methods of Determination of Numerical Electronegativities," *Russ. J. Phys. Chem.*, **1963**, *37*, 1070-1072.

11. V. P. Spiridonov, M. Tatevskii, "The Concept of Electronegativity. V. The Meaning of Electronegativities and their Use for Calculating the Properties of Molecules" *Russ. J. Phys. Chem.*, **1963**, *37*, 1177-1179.

12. C. K Jørgensen, *Inorganic Complexes*, Academic Press: London, 1963, p. 4.

13. L. H. Ahrens, *Ionization Potentials: Some Variations, Implications and Applications*, Pergamon Press: Oxford, 1983.

14. H-A. Lehmann, S. Bähr, "Über die Zussamenhang zwischen Elektronegativitäten

und Rumpflandungszahlen," in W. Triebs, Ed., *Elekronentheorie der Homöopolaren Bindung*, Akademie Verlag: Berlin, 1956, pp. 20-28.

15. H-A. Lehmann, S. Bähr, "Über die Abschätzung von Bindungspolaritäten mit Hilfe von Rumpfladungszahlen und ihre Beziehungen zu den Elektronegativitäten nach Pauling," *Z. anorg. Chem.*, **1956**, *287*, 1-11.

16. A. I. Martynov, S. S. Batsanov, "A New Approach to the Determination of the Electronegativity of Atoms," *Russ. J. Inorg. Chem.*, **1980**, *25*, 1737-1739.

17. W. B. Jensen, *Lectures on Descriptive Inorganic Chemistry and the Periodic Law,* University of Cincinnati: Cincinnati, OH,1995, Chapter 7. The values given here were first reported by the author in a seminar given at the University of Wisconsin in 1980.

18. L. C. Allen, "Electronegativity is the Average One-Electron Energy of the Valence-Shell Electrons in the Ground-State Free Atom." *J. Am. Chem. Soc.*, **1989**, *111*, 9003-9014.

19. L. C. Allen, "Electronegativity and the Periodic Table," in P. von Ragué Schleyer, Ed., *Encyclopedia of Computational Chemistry*, Vol. 2, Wiley: New York, NY, 1998, pp. 835-852. EN_S values in units of eV are given for the main-block and transition-block elements in Tables 2 and 3 of this reference.

Appendix II
AB Semiconductors

Compound	*ΔFEN*	FEN_{av}	*Structure*[a]	E_g/eV	*Color*
Group 1/7 - Group 7/1					
LiF	0.95	0.53	S	13.6	colorless
LiCl	0.69	0.40	S	9.4	colorless
LiBr	0.64	0.37	S	7.6	colorless
LiI	0.57	0.34	S	~ 6	colorless
NaF	0.96	0.52	S	11.6	colorless
NaCl	0.70	0.39	S	8.5	colorless
NaBr	0.65	0.37	S	7/5	colorless
NaI	0.58	0.33	S	~ 6	colorless
KF	0.98	0.51	S	10.7	colorless
KCl	0.72	0.38	S	8.4	colorless
KBr	0.67	0.36	S	7.4	colorless
KI	0.60	0.32	S	6.0	colorless
RbF	0.99	0.51	S	10.3	colorless
RbCl	0.73	0.38	S	8.2	colorless
RbBr	0.68	0.35	S	7.4	colorless
RbI	0.61	0.32	S	6.1	colorless
CsF	1.00	0.00	S	9.9	colorless
CsCl	0.74	0.37	F	8.3	colorless
CsBr	0.69	0.35	F	7.3	colorless
CsI	0.62	0.31	F	6.2	colorless

Group 2/6 - Group 6/2

BeO	0.61	0.55	D	--	colorless
BeS	0.38	0.43	D		
BeSe	0.36	0.42	D		
BeTe	0.31	0.40	D		
MgO	0.66	0.52	S	7.3	colorless
MgS	0.43	0.41	S		red-brown
MgSe	0.41	0.40	S		light gray
MgTe	0.36	0.37	D		colorless
CaO	0.72	0.49	S	6.0	colorless
CaS	0.49	0.38	S		colorless
CaSe	0.47	0.37	S		
CaTe	0.42	0.34	S		
SrO	0.73	0.49	S		colorless
SrS	0.50	0.37	S		light gray
SrSe	0.48	0.36	S		colorless
SrTe	0.43	0.34	S		colorless
BaO	0.75	0.48	S	6.0	colorless
BaS	0.52	0.36	S		colorless
BaSe	0.50	0.35	S		colorless
BaTe	0.45	0.33	S		light yellow
ZnO	0.61	0.55	S	3.2	colorless
ZnS	0.38	0.43	D	3.54	colorless
ZnSe	0.36	0.42	D	2.58	yellow-red
ZnTe	0.31	0.40	D	2.26	red
CdO	0.63	0.54	D	2.50	brown
CdS	0.40	0.42	D	2.58	yellow-orange
CdSe	0.38	0.41	D	1.74	red
CdTe	0.33	0.39	D	1.44	black
HgO	0.60	0.55	L		red

HgS	0.34	0.44	D	2.10	red
HgSe	0.35	0.43	D	2.10	gray
HgTe	0.30	0.40	D		

Group 3/5 - Group 5/3

BN	0.32	0.54	D	4.6	colorless
BP	0.14	0.45	D	3.1	red
BAs	0.12	0.44	D	1.5	
BSb	0.08	0.42	U		
AlN	0.41	0.50	D	6.02	colorless
AlP	0.23	0.41	D	2.45	green-yellow
AlAs	0.21	0.40	D	2.16	orange
AlSb	0.17	0.38	D	1.60	
GaN	0.39	0.51	D	3.34	gray
GaP	0.21	0.42	D	2/24	yellow
GaAs	0.19	0.41	D	1.35	gray
GaSb	0.15	0.39	D	0.67	
InN	0.42	0.49	D	2.00	
InP	0.24	0.40	D	1.27	gray
InAs	0.22	0.39	D	0.36	gray
InSb	0.18	0.37	D	0.16	black

Group 4/4 - Group 4/4

CC	0.00	0.54	D	5.7	colorless
SiSi	0.00	0.40	D	1.11	gray
GeGe	0.00	0.40	D	0.67	gray
Sn-Sn	0.00	0.38	D	0.08	gray
SiC	0.14	0.47	D	2.40	black

[a] S = 6/6 Sodium Chloride, D = 4/4 Diamond, L = 2/2 Linear, F = 8/8 Fluorite, U = Unknown

Printed in the United States
by Baker & Taylor Publisher Services